Unified low-power design flow for data-dominated multi-media and telecom applications

Unified low-power design flow for data-dominated multi-media and telecom applications

Based on selected partner contributions of the European Low Power Initiative for Electronic System Design of the European Community ESPRIT4 programme

Edited by

Francky Catthoor
IMEC, Leuven, Belgium

Series Editors

René van Leeuwen
Reinder Nouta
Alexander de Graaf
TUDelft/DIMES, Delft, The Netherlands

Springer-Science+Business Media, B.V.

A CIP Catalogue record is available from the Library of Congress

ISBN 978-1-4419-5000-0 ISBN 978-1-4757-3182-8 (eBook)
DOI 10.1007/978-1-4757-3182-8

Printed on acid-free paper

Originally published by Kluwer Academic Publishers, Boston in 2000.
Softcover reprint of the hardcover 1st edition 2000

Contents

List of Figures

List of Tables

Contributing Authors

Francky Catthoor received the Eng. degree and a Ph.D. in El. Eng. from the K.U.Leuven, Belgium in 1982 and 1987 respectively. Since 1987, he has headed research domains in the area of architectural and synthesis methodologies, within the DESICS (formerly VSDM) division at IMEC. He is part-time professor at the EE department of the K.U.Leuven since 1989. His current research activities mainly belong to the field of application-specific architecture design methods and system-level exploration, with emphasis on memory and global data transfers. Both customized architectures and programmable (parallel) multimedia processors are targeted.

Erik Brockmeyer received his degree in Electrical Engineering in 1998 from the University of Eindhoven, the Netherlands. In 1997 he did his master thesis on MPEG-4 in the System Exploration for Memory and Power (SEMP) group. This group is part of the DESICS division of the Interuniversity Micro Electronics Center (IMEC). Currently he is working at IMEC in the Embedded Systems (EMSYS) group, in the field of Data Transfer and Storage Exploration (DTSE), with emphasis on storage cycle budget distribution.

Paul Lippens was born in Aardenburg, The Netherlands, in 1962. He received the electrical engineering degree from the Eindhoven University of Technology in 1986 (with honors). He worked as a member of the 'Digital VLSI Design' group at Philips Research Laboratories in Eindhoven. There, he was involved in architectural-level synthesis for high-throughput applications and has authored and co-authored several publications in this area. Presently, he is working on heterogeneous multi-processor architectures and hardware-software co-design in the 'Embedded Systems Architectures on Silicon' group at Philips Research Labs in Eindhoven.

Roger Woods is a senior lecturer in the Queen's University of Belfast. His main research interests include VLSI architectures (particularly low power), architectural synthesis of VLSI signal processing chips and dynamically reconfigurable logic. He has worked closely with industry and has developed chips that are being commercially exploited by Mitel, BBC and Snell and Wilcox. He holds two patents and has published over 70 papers in international journals and conferences. He will act as general chair for the Field Programmable Logic conference in Belfast in 2001 and is involved in the program committees of the IEEE Symposium on FPGA-based Custom Computing Machines, International conference on Field Programmable Logic, VLSI'99 and the IEEE conference on Application Specific Array Processors.

Nikos D. Zervas received a Diploma degree in Electrical and Computer Engineering in 1997, from the Department of Electrical and Computer Engineering at the University of Patras, Greece. He has been at the same department and university since 1998 working on his Ph.D.. His research interests are in the area of high-level, power optimization techniques for multimedia and telecommunication applications. He has received an award in the context of Low-Power Design Contest of 2000 IEEE Computer Elements Mesa Workshop. Mr. Zervas is a member of the Greek Technical Chamber and a student member of the IEEE.

Konstantinos Masselos received a first degree in Electrical Engineering from University of Patras, Greece in 1994, an MSc degree in VLSI Systems Engineering from University of Manchester Institute of Science and Technology (UMIST), United Kingdom in 1996, and a Ph.D. degree from University of Patras, Greece in 2000. For his MSc studies he was awarded a UMIST graduate research scholarship. His research concerns system level low power design methodologies for multimedia applications realized on different architectural platforms and includes data storage and transfer optimization, design of low complexity algorithms and efficient implementations for image and video processing, high level synthesis and compilation techniques. He is associated as a visiting researcher to the System Exploration for Memory and Power (SEMP) group of the VLSI Systems Design Methodologies (VSDM) division of the Inter-university Micro Electronics Centre (IMEC) in Leuven, Belgium, where he is involved in research related to the ACROPOLIS multimedia pre-compiler. He is a member of the IEEE and of the Technical Chamber of Greece.

Foreword

This book is the first in a series on novel low power design architectures, methods and design practices. It results from of a large European project started in 1997, whose goal is to promote the further development and the faster and wider industrial use of advanced design methods for reducing the power consumption of electronic systems.

Low power design became crucial with the wide spread of portable information and communication terminals, where a small battery has to last for a long period. High performance electronics, in addition, suffers from a permanent increase of the dissipated power per square millimetre of silicon, due to the increasing clock-rates, which causes cooling and reliability problems or otherwise limits the performance.

The European Union's Information Technologies Programme 'Esprit' did therefore launch a 'Pilot action for Low Power Design', which eventually grew to 19 R&D projects and one coordination project, with an overall budget of 14 million Euro. It is meanwhile known as European Low Power Initiative for Electronic System Design (ESD-LPD) and will be completed by the end of 2001. It involves 30 major European companies and 20 well-known institutes. The R&D projects aims to develop or demonstrate new design methods for power reduction, while the coordination project takes care that the methods, experiences and results are properly documented and publicised.

The initiative addresses low power design at various levels. This includes system and algorithmic level, instruction set processor level, custom processor level, RT-level, gate level, circuit level and layout level. It covers data dominated and control dominated as well as asynchronous architectures. 10 projects deal mainly with digital, 7 with analogue and mixed-signal, and 2 with software related aspects. The principal application areas are communication, medical equipment and e-commerce devices.

The following list describes the objectives of the 20 projects. It is sorted by decreasing funding budget.

CRAFT CMOS Radio Frequency Circuit Design for Wireless Application

- Advanced CMOS RF circuit design including blocks such as LNA, down converter mixers & phase shifters, oscillator and frequency synthesiser, integrated filters delta sigma conversion, power amplifier
- Development of novel models for active and passive devices as well as fine-tuning and validation based on first silicon fabricates
- Analysis and specification of sophisticated architectures to meet in particular low power single chip implementation

PAPRICA Power and Part Count Reduction Innovative Communication Architecture

- Feasibility assessment of DQIF, through physical design and characterisation of the core blocks
- Low-power RF design techniques in standard CMOS digital process
- RF design tools and framework; PAPRICA Design Kit.
- Demonstration of a practical implementation of a specific application

MELOPAS Methodology for Low Power Asic design

- To develop a methodology to evaluate the power consumption of a complex ASIC early on in the design flow
- To develop a hardware/software co-simulation tool
- To quickly achieve a drastic reduction on the power consumption of electronic equipment

TARDIS Technical Coordination and Dissemination

- To organise the communication between design experiments and to exploit their potential synergy
- To guide the capturing of methods and experiences gained in the design experiments
- To organise and promote the wider dissemination and use of the gathered design know-how and experience

LUCS Low Power Ultrasound Chip Set.

- Design methodology on low power ADC, memory and circuit design
- Prototype demonstration of a handheld medical ultrasound scanner

ALPINS Analogue Low Power Design for Communications Systems

- Low-voltage voice band smoothing filters and analogue-to-digital and digital-to-analogue converters for an analogue front-end circuit of a DECT system
- High linear transconductor-capacitor (gm-C) filter for GSM Analogue Interface Circuit operating at supply voltages as low as 2.5V
- Formal verification tools, which will be implemented in the industrial partners design environment. These tools support the complete design process from system level down to transistor level

SALOMON System-level analogue-digital trade-off analysis for low power

- A general top-down design flow for mixed-signal telecom ASICs
- High-level models of analogue and digital blocks and power estimators for these blocks
- A prototype implementation of the design flow with particular software tools to demonstrate the general design flow

DESCALE Design Experiment on a Smart Card Application for Low Energy

- The application of highly innovative handshake technology
- Aiming at some 3 to 5 times less power and some 10 times smaller peak currents compared to synchronously operated solutions

SUPREGE A low power SUPerREGEnerative transceiver for wireless data transmission at short distances

- Design trade-offs and optimisation of the micro power receiver/transmitter as a function of various parameters (power consumption, area, bandwidth, sensitivity, etc)
- Modulation / demodulation and interface with data transmission systems
- Realisation of the integrated micro power receiver / transmitter based on the super-regeneration principle

PREST Power REduction for System Technologies

- Survey of contemporary Low Power Design techniques and commercial power analysis software tools
- Investigation of architectural and algorithmic design techniques with a power consumption comparison
- Investigation of Asynchronous design techniques and Arithmetic styles
- Set-up and assessment of a low power design flow
- Fabrication and characterisation of a Viterbi demonstrator to assess the most promising power reduction techniques

DABLP Low Power Exploration for Mapping DAB Applications to Multi-Processors

- A DAB channel decoder architecture with reduced power consumption
- Refined and extended ATOMIUM methodology and supporting tools

COSAFE Low Power Hardware-Software Co-Design for Safety-Critical Applications

- The development of strategies for power efficient assignment of safety critical mechanisms to hardware or software
- The design and implementation of a low-power, safety-critical ASIP, which realises the control unit of a portable infusion, pump system

AMIED Asynchronous Low-Power Methodology and Implementation of an Encryption/Decryption System

- Implementation of the IDEA encryption/decryption method with drastically reduced power consumption
- Advanced low power design flow with emphasis on algorithm and architecture optimisations
- Industrial demonstration of the asynchronous design methodology based on commercial tools

LPGD A Low-Power Design Methodology/Flow and its Application to the Implementation of a DCS1800-GSM/DECT Modulator/Demodulator

- To complete the development of a top-down, low power design methodology/flow for DSP applications
- To demonstrate the methods at the example of an integrated GFSK/GMSK Modulator-Demodulator (MODEM) for DCS1800-GSM/DECT applications

SOFLOPO Low Power Software Development for Embedded Applications

- Develop techniques and guidelines for mapping a specific algorithm code onto appropriate instruction subsets
- Integrate these techniques into software for the power-conscious ARM-RISC and DSP code optimisation

I-MODE Low Power RF to Base band Interface for Multi-Mode Portable Phone

- To raise the level of integration in a DECT/DCS1800 transceiver, by implementing the necessary analogue base band low-pass filters and data converters in CMOS technology using low power techniques

COOL-LOGOS Power Reduction through the Use of Local don't Care Conditions and Global Gate Resizing Techniques: An Experimental Evaluation.

- To apply the developed low power design techniques to the existing 24-bit DSP, which is already fabricated
- To assess the merit of the new techniques using experimental silicon through comparisons of the projected power reduction (in simulation) and actually measured reduction of new DSP; assessment of the commercial impact

LOVO Low Output VOltage DC/DC converters for low power applications

- Development of technical solutions for the power supplies of advanced low power systems, comprising the following topics
- New methods for synchronous rectification for very low output voltage power converters

PCBIT Low Power ISDN Interface for Portable PC's

- Design of a PC-Card board that implements the PCBIT interface
- Integrate levels 1 and 2 of the communication protocol in a single ASIC
- Incorporate power management techniques in the ASIC design:
  - system level: shutdown of idle modules in the circuit
  - gate level: precomputation, gated-clock FSMs

COLOPODS Design of a Cochlear Hearing Aid Low-Power DSP System

- Selection of a future oriented low-power technology enabling future power reduction through integration of analogue modules
- Design of a speech processor IC yielding a power reduction of 90% compared to the 3.3 Volt implementation

The low power design projects have achieved the following results:

- Projects, who have designed a prototype chip, can demonstrate a power reduction of 10 to 30 percent.
- New low power design libraries have been developed.
- New proven low power RF architectures are now available.
- New smaller and lighter mobile equipment is developed.

Instead of running a number of Esprit projects at the same time independently of each other, during this pilot action the projects have collaborated strongly. This is achieved mostly by the novelty of this action, which is the presence and role of the coordinator: DIMES - the Delft Institute of Microelectronics and Submicron-technology, located in Delft, the Netherlands (http://www.dimes.tudelft.nl). The task of the coordinator is to co-ordinate, facilitate, and organize:

- The information exchange between projects.
- The systematic documentation of methods and experiences.
- The publication and the wider dissemination to the public.

The most important achievements, credited to the presence of the coordinator are:

- New personnel contacts have been made, and as a consequence the resulting synergy between partners resulted in better and faster developments.
- The organization of low power design workshops, special sessions at conferences, and a low power design web site, http://www.esdlpd.dimes.tudelft.nl. At this site all public reports of the projects can be found and all kind of information about the initiative itself.
- The used design methodology, design methods and/or design experience are disclosed, are well documented and available.

  Based on the work of the projects, in cooperation with the projects, the publication of a low power design book series is planned. Written by members of the projects this series of books on low power design will disseminate novel design methodologies and design experiences, which were obtained during the runtime of the European Low Power Initiative for Electronic System Design, to the general public.

In conclusion, the major contribution of this project cluster is that, except the already mentioned technical achievements, the introduction of novel knowledge on low power design methods into the mainstream development processes is accelerated.

We would like to thank all project partners from all the different companies and organizations who make the Low Power Initiative a success.

Rene van Leuken, Reinder Nouta, Alexander de Graaf, Delft, June 2000

INTRODUCTION

This edited book brings together several low-power oriented design flows which are being developed in several European research groups into a more abstract but unifying design flow oriented to data-dominated multi-media and telecom applications. From this unified flow, a particular design flow can then still be instantiated for a given application in the target domain by leaving out the non-required stages/steps and by selecting the appropriate technique for all remaining (sub)steps. We hope that this material can be of use for many different contexts, especially in introducing a more systematic and overall design flow, at higher abstraction levels than what is currently commercially supported.

The material in this document is based on the background of the different partners involved in the DAB-LP, PREST, LPGD and COLOPODS projects of the European Low Power Initiative for Electronic System Design. They have been working for many years on low-power design issues and hence the ideas are directly or indirectly influenced by our many colleagues (past and current). Also interesting discussions with colleagues in other research groups, both academic and industrial, have influenced the final result.

The goal of the low power system design work within these projects has been to contribute design methodologies and application designs which address the low power design trajectory from *real behaviour* down to the RT-level *structural specification* of the system. In order to provide complete support for this synthesis trajectory, many design problems must be tackled. We do not claim to cover the complete path, but we do believe we have contributed to the solution of a number of the most crucial problems in the domains of design methodologies and exploration support. We therefore expect this book to be of interest in academia; not for detailed descriptions of the research results — these have been published elsewhere as indicated in the extensive bibliographies of the chapters — but for the overview of the field and a view on the many important but less widely known issues which must be addressed to arrive at industrially relevant results.

All the projects of the European Low Power Initiative have also been application-driven from the start, and the book is intended to reflect this fact. The real-life applications that have driven the research are described, and the impact of their characteristics on the methodologies is assessed. We therefore believe that the book will be of interest to senior design engineers and CAD managers in industry, who wish either to anticipate the evolution of commercially available low power design methodologies tools over the next few years, or to make use of the concepts in their own research and development.

It has been a pleasure for me to work on this book project. The coordination of the book has meant many hours on the phone, answering e-mails, and behind computer keyboards. However, during this time I have had the opportunity to obtain more in-depth knowledge about many new issues in low power system design, and to develop close connections with excellent people at each of the contributing partner sites.

The authors of the different chapters in this book and myself would like to use this opportunity to thank the many people who have helped make these projects successful, and to express our appreciation of their contributions:

The authors of chapter 2 and 3 wish to thank their colleagues at IMEC and especially in the System Exploration for Memory and Power (SEMP) domain for the valuable input and discussions on the initial versions of the material which ultimately got collected in this book.

The authors of chapter 5 wish to wish to thank their colleagues in the School of Electrical and Electronic Engineering at Queen's University and input from their partners in the PREST project particularly Steve Furber, Peter Riocreux, Peter Ivey, Luke Seed, George Abouyannis, Keith Lyons and Gary Coulson.

The authors of chapter 6 wish to thank Prof. G. Kalivas and A. Thanailakis for their encouragement, support and the fruitful conversations. The authors also wish to thank Drs. C. Dre and K. Katis from INTRACOM S.A. for sharing their industrial experience and their colleagues A. Tzimas, E. Metaxakis, X. Mizas and V. Mardiris for their valuable help.

The authors of chapter 7 wish to acknowledge the discussions with their colleagues in the VLSI Design Laboratory of the Department of Electrical and Computer Engineering of University of Patras and especially with Panagiotis Merakos, Spyros Theoharis and Professor Thanos Stouraitis.

All of us hope that the reader will find the book useful and enjoyable, and that the results presented will contribute to the continued progress of the field of system-level low power design methodologies and techniques.

Francky Catthoor, Leuven, April 2000

1 MOTIVATION, CONTEXT AND OBJECTIVES

Francky Catthoor

IMEC, Leuven, Belgium

catthoor@imec.be

Rene van Leuken, Reinder Nouta, Alexander de Graaf

TUDelft/DIMES, Delft, Netherlands

esdlpd@dimes.tudelft.nl

A number of considerations drive the electronic industry to try to reduce the power consumption of electronic systems. Among the more important ones are the following.

- **cooling and system packaging considerations:** Electronic systems are not 100 percent efficient at converting their input power (from the mains or a battery) into useful power (sound, picture, etc). Some of the power is lost in the system and is converted to heat, which has to be removed by cooling otherwise sensitive electronic components could overheat and possibly fail. In some cases forced air-cooling may be required. This all adds to the system cost. Therefore low power systems need less cooling since their input power requirement is lower. This in turn means that the power supply unit can also be smaller and therefore cheaper.

- **reliability:** Low power systems enable the silicon chips to run at lower temperatures and are therefore more reliable and less likely to malfunction. This is because the electro-migration at lower temperatures is less. Another contributing factor to improved reliability is, that the failure mechanisms inside the chips are temperature dependent so the lower the operating temperature the less failures that result.

F. Catthoor (ed.),
Unified Low-power Design Flow for Data-dominated Multi-media and Telecom Applications, 1-6.

- **power/frequency considerations:** Electronic systems are required to work at higher and higher frequencies in order to give faster operation and higher performance. The higher the frequency of operation the higher the power will be unless steps are introduced in the design to reduce the power to acceptable levels.

- **battery-life:** The development of the mobile phone illustrates the need for low power very well. Early models were large and bulky (mainly because of the large batteries required), and required battery renewal/ re-charging almost on a daily basis due mainly to the power-hungry analogue circuitry used in them. The latest models are designed for low power, are very small and light, have long operating life, have advanced features such as the ability to access the Internet, and are cheap and affordable to all.

- **new applications:** Having a low power design methodology available, may result in a company having competitive advantages from the ability to bring new applications and new products to market, that otherwise not have been possible.

Given the above described reasons it is necessary to take power dissipation considerations into account during the design process, besides of the "classical" performance parameters (delay and area). This makes the design an even more complex task, since the designer is now responsible for finding an optimum in the three dimensional delay-power-area design space.

Many different design flows (or partial flows) exist in the system and architecture design literature (see e.g. [4, 5, 15]), usually intended also for different application domains. Many of these flows have resemblances but it is usually difficult to pinpoint where the corresponding parts are and how you should apply them to create a low-power realisation for a particular application. This is for a large part due to incompatible terminology, the presence of large gaps in the design flows and the lack of unambiguous information which obscure the global picture. In order to allow more "reuse" of knowledge and information it is vital to overcome this problem. For this purpose, a common terminology basis is a first step. In addition, the principles of a unified low-power oriented design flow should be provided in sufficient detail so that designers can instantiate and apply this for their particular applications. Obviously, this would be a very ambitious objective for the entire range of application and architecture targets.

In this book, we will try to define a unified system design terminology and low-power system design flow for a quite broad target domain of data- and arithmetic-dominated multi-media applications. It will not contain the very details of each individual step because these are described in other documents (see appropriate references in the subsequent chapters). The target application domain of the system-level low power approach advocated by us is advanced real-time multi-media and information processing (RMP) systems, such as consumer multi-media electronics (including video, image, audio and speech processing) and personal communication systems. These applications belong to the class of real-time embedded systems involving a combination of complex data- and control-flow, where also complex data types are manipulated and transferred. Most of these applications are compact and portable devices, putting stringent constraints on the degree of integration (i.e. chip area) and on their power consumption. These systems are also extremely *heterogeneous in nature* and combine high performance data processing (e.g. data processing on transmission data input) as well as slow rate control processing (e.g. system control functions), synchronous as well as asynchronous parts, analog versus digital. Time-to-market has become a critical factor in the design phase, so the use of a formalized design methodology, partly supported by tools, is crucial. Finally, these systems are subjected to stringent real-time constraints (both hard and soft deadlines are present), complicating their implementation considerably.

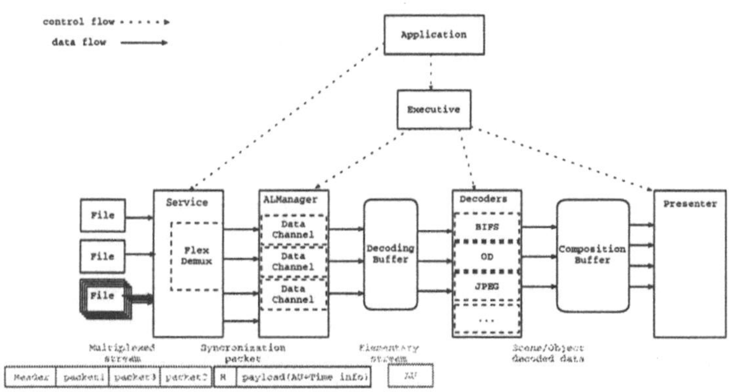

Figure 1.1 System layer of MPEG4 IM1 player, with several concurrent modules and complex dynamic control constructs.

A major driver for this type of research are applications like the new MPEG4 standard (see fig. 1.1), which involves a massive amount of specification code (more than 100K lines of high-level C++ code) and which combines video, synthetic images, audio and speech modules with a complex dynamic system protocol [2].

Another demonstrator is given in fig. 1.2, depicting a mobile terminal for satellite voice and data communication [1].

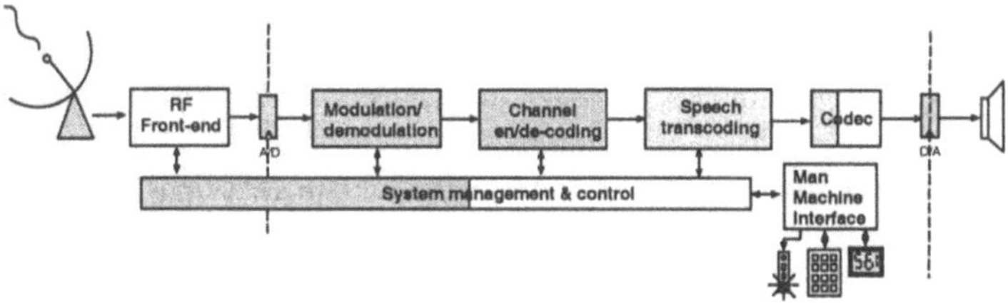

Figure 1.2 A mobile satellite terminal.

The target architectures are very broad also. Today, a new heterogeneous architectural design paradigm is emerging (see fig. 1.3), usually called a "platform" or "systems-on-a-chip", including one or more programmable components, either general-purpose or DSP processors, cores or ASIPs (application-specific instruction-set processor), augmented with some specialized data paths or co-processors (accelerators). The programmable components run software components, being slow to medium speed algorithms, while time-critical parts are executed on dedicated hardware accelerators. Next to these programmable components, they contain large amounts of memory organizations (DRAM, SRAM, FIFOs, ...), and their supporting address calculation units (ACUs) or memory management units (MMUs), I/O peripherals, interface logic and more and more analog blocks as well.

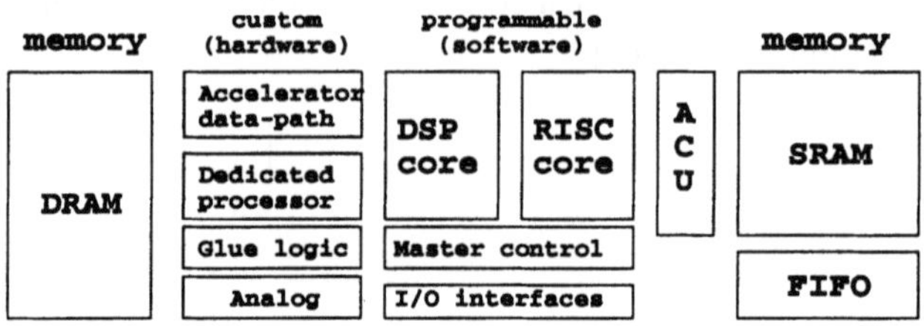

Figure 1.3 A typical heterogeneous implementation platform for embedded multi-media systems.

The abstraction levels at which we focus our attention are the ones between a truly algorithmic specification, usually augmented with some constraints on the system architecture platform to be targeted, and the processor-level system architecture. The latter still has to be synthesized into a circuit for the hardware partition and the software, which is usually represented in a language like C, has to be compiled on the assigned target processors.

In order to arrive at a broadly applicable design flow in this important part of the system design trajectory, and to really use a common terminology, there is a need to have the same view of an underlying system design meta flow. A proposal for the scope of this book is provided in chapter 2, focused on the low power aspects of the task- and processor-level abstraction stages and illustrated by an application from the multi-media domain.

One of the key factors in modern multi-media and telecom applications involves the manipulation (transfer and storage) of the complex data types. We have shown that these issues should be dealt with very early in the system design trajectory, and the resulting design methodology is the topic of chapter 3. The methodology minimizes in a systematic way the memory storage and transfers in order to reduce the power (and area) for data dominated applications in customizable memory organisations.

In order to steer such a system design methodology, sufficient information about the power budget distribution should be available. At high levels, only estimates of these data are feasible. The material in chapter 4 forms a clear contribution to raise the abstraction level at which sufficiently accurate power estimates can be obtained. The techniques are illustrated mainly on the exploration for data transfer and storage issues of chapter 3.

The remainder of the book mainly focuses on different issues in the design of custom processor architectures. Chapter 5 addresses high-throughput signal processing applications that are both arithmetic- and data-dominated, and where the performance capabilities of regular customized processor solutions are particularly attractive. The designs resulting from the tuned approach proposed there, are characterized by high area utilization, high levels of locality (preserving power) and efficient memory utilization.

Chapter 6 addresses run-time power management, which is one of the most efficient low-power strategies, applicable at various levels of abstraction of our unified design flow. Two run-time power management techniques applicable for the design of low and medium bit-rate digital modems are introduced. Specifically, an event-driven power management methodology and a dynamic power management technique based on frequency scaling, are presented for custom processors.

Finally, chapter 7 focuses on techniques for the power-efficient data-path synthesis of sum-of-products computations. Heuristics for the instruction-level scheduling and assignment steps are described.

References

[1] C. V. Himbeeck, I. Deman, M. Dothey, L. Philips, I. Bolsens, A. Rabaeijs, B. Vanhoof, J. Vanhoof and H. D. Man. – Silicon synthesis of a flexible CDMA/QPSK mobile communication modem. *DSP Applications*, Jan. 1994.

[2] T.Sikora, "The MPEG-4 video standard verification model", *IEEE Trans. on Circuits and Systems for Video Technology*, Vol.7, No.1, pp.19–31, Feb. 1997.

2 UNIFIED META-FLOW SUMMARY FOR LOW-POWER DATA-DOMINATED APPLICATIONS

Francky Catthoor and Erik Brockmeyer

IMEC, Leuven, Belgium

catthoor@imec.be,brockmey@imec.be

In order to arrive at a broadly applicable design flow in this important part of the system design trajectory, and to really use a common terminology, there is a need to have the same view of an underlying system design meta flow. The basic high-level synthesis terminology proposed in [20, 21, 27] was useful but not sufficient to cover the rapidly advancing research. At IMEC, a unified meta flow bridging an even larger span of abstraction levels has been proposed (see overview in [8]). The resulting proposal is (to our knowledge) the first attempt to obtain unification over such a broad scope.

In this chapter, the relevant parts for our purpose are summarized in chapter 2, focused on the low power aspects of the task- and processor-level abstraction stages and illustrated by an application from the multi-media domain.

2.1 MAIN PRINCIPLES

Since this flow targets a very broad application domain, not all of its (sub)steps are important for every type of design. In fact, it represents a *meta* flow which is to be instantiated by omitting those (sub)steps which are irrelevant from the viewpoint of the specific type of application to be designed. Moreover, depending on the type of application domain which is considered, a specific step in the meta flow will be instantiated in a partly different way[1].

The global unified meta flow presents a single overall framework which is surprisingly simple but certainly not trivial. The strict separation between different abstraction layers allows the designer to deal with the issues on each level more clearly and in a more decoupled fashion, and provides feedback as early as possible, avoiding large scope iterations through the entire design process. Also the historical distinction

F. Catthoor (ed.),
Unified Low-power Design Flow for Data-dominated Multi-media and Telecom Applications, 7-23.

between "hardware" and "software" is discarded, thereby removing the artificial and inefficient boundaries between much existing work.

2.2 OBJECTIVES

This meta flow attempts to meet the following objectives:

1. Achieve a common terminology, as stated above

2. Simplify identification of related work and comparison. In principle, a single common database of references related to each of the "hooks" in the unified flow can be envisioned. When someone is interested in a (sub)step, all related info is then accessible in a simple way.

3. Establish a framework to identify "holes" in a given (new) flow. In most application domains there will be many (sub)steps in the unified flow which are not relevant or which can be merged with other (sub)steps. However, when this issue is incorporated, all the remaining (sub)steps should be somehow present in the final solution and hence it is better to take this into account from the start of the system design support project.

4. Remove redundancy and unnecessary overlap. Stimulates knowledge reuse and in the end also software reuse.

2.3 CHARACTERISTICS

The characteristics of the proposed meta flow are:

1. It represents a unified meta-flow from which many instantiated flows can be derived.

2. In principle, it is independent of the type of system design, as long as one remains within an electronic context.

3. It is **not** a single detailed design flow focused on a specific target application domain. So even given this unified flow, the amount of research effort to arrive at a customized and efficient flow for a given target domain is still large. The coupling between efficient synthesis support and target applications still remains valid.

4. The dependencies between the (sub)steps should be obeyed in any "logical" instantiated flow, but it is always allowed to group (sub)steps and to reorder any pair of (sub)steps which does not exhibit a dependence (just like many schedules/allocations can be derived from a single data-flow graph).

5. It is definitely **not** a tool flow because then also the CAD algorithms should be incorporated and these are even more target domain specific.

6. It also does not incorporate design data models (e.g. data-flow vs control-flow) because that is not necessary and would hamper the unification. This abstraction of the underlying data model is a crucial enabling issue.

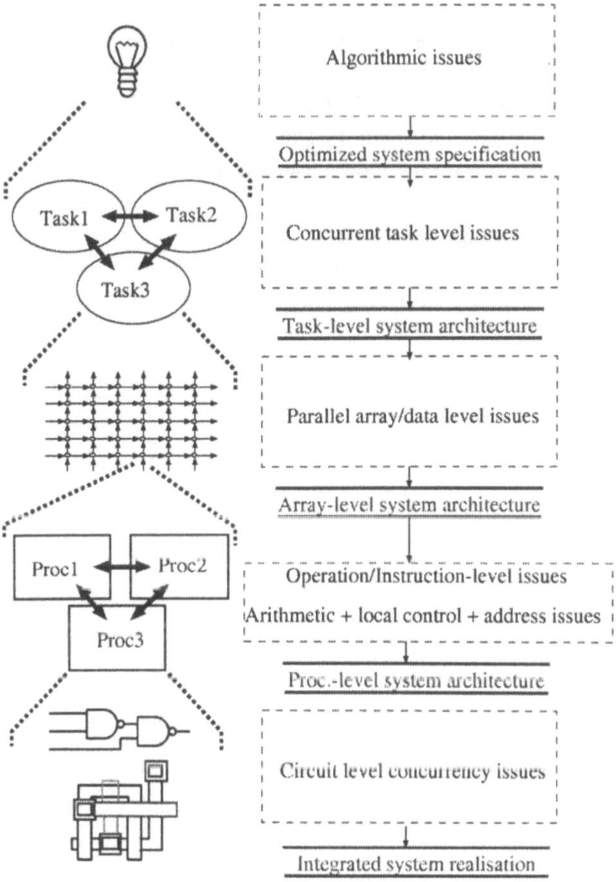

Figure 2.1 Basic structure of the unified meta flow

2.4 GLOBAL VIEW AND MORE DETAILED PRINCIPLES

A global view on the proposal is provided in figure 2.1. In this unified flow, we can identify a number of abstraction layers. Each group a number of (sub)steps. The major stages in the system design trajectory are related to the algorithm design, the task-level concurrency, the data parallelisation into regular arrays, the instruction-level concurrency and the circuit-level concurrency. At each concurrency level, both the data transfer and storage issues (in the data transfer and storage exploration (DTSE) steps) and the concurrency management issues are addressed[2]. The DTSE steps are all related to the handling of complex data types such as indexed signals (arrays) or dynamically created lists (tables) (see figure 2.2). It involves global data analysis (for dependencies, accesses and memory size estimates), global data- and control-flow transformations (to remove redundancy and data-flow bottle-necks respectively to increase access regularity and locality), data reuse decisions (related to the exploitation of memory hierarchy), storage cycle budget distribution (to meet timing constraints), memory allocation and assignment (for background memories) and in-place mapping (to reuse data space for arrays with a limited, partially overlapping life-time).

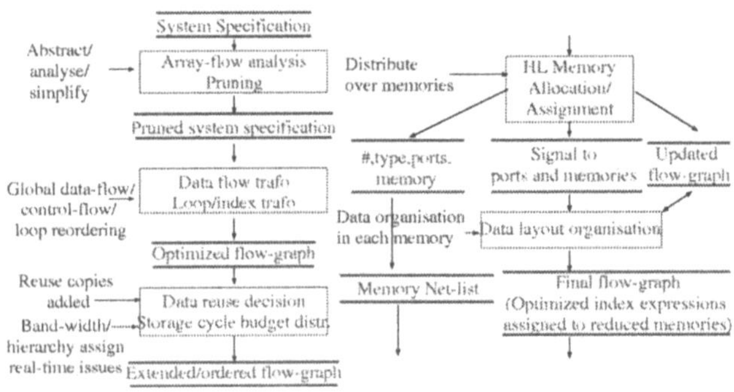

Figure 2.2 DTSE issues and constraints.

In contrast, the concurrency management steps are related to concurrency extraction, concurrent execution (within a time budget), the synchronisation (ordering), the resource allocation (within a time budget) and the interface refinement between the concurrent entities (figure 2.3). It has to be stressed that the DTSE stages are always ordered before their concurrency stage counterpart at the 3 main abstraction layers. This is necessary because they provide vital information on the communication ordering constraints and costs issues related to the background memory storage/transfer of the most important internal signals in each task. Reversing the order would also mean that the concurrency stage provides strong restrictions on the data communication ordering, which would be inconsistent with the reason for having a separate DTSE stage [6, 7].

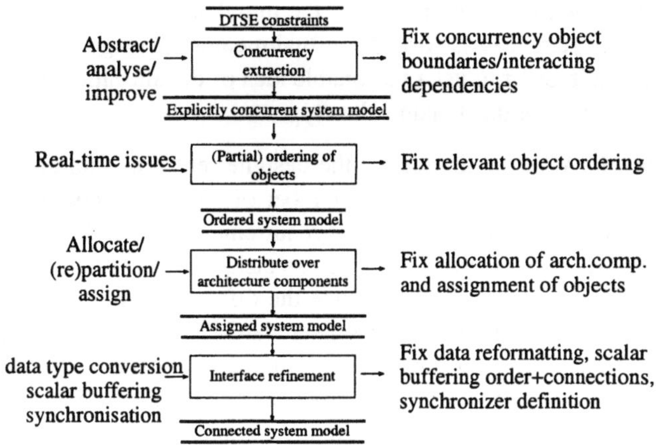

Figure 2.3 Concurrency issues and constraints

2.5 OVERVIEW OF MAIN STAGES AND STEPS

A rundown of the various "stages" in the meta flow is given now with emphasis on the instruction/operation-level abstraction layer, because that is also the main focus of the remainder of this book. The idea is to briefly introduce the issues that are dealt with in the different steps when we open some of the boxes in figure 2.1. It has to be stressed that several extra layers of refinement are present in the complete flow but in this book we cannot provide a more detailed and formal explanation. Some references to additional material have been added though.

1. **algorithm level:** Early stages in a design flow concentrate on generating a correct functional specification of the design.

2. **data type refinement:** Since the specification of a system is described in terms of *abstract data types*, e.g. real (infinite precision) numbers, lists and arrays, it is necessary to refine these data types in terms of basic (implementable) data structures. This applies for the "normal" data as well as for the "parameters". Examples of this include the quantization of filter coefficients for the parameters, and the optimization of the dynamic range for data under given constraints on the quantisation/overflow/limit cycle effects which can be tolerated (see e.g. [24]). For the dynamically created data, also dynamic memory management has to be dealt with. This includes both an abstract data type refinement (ADT) step where e.g. sets with access keys are refined into linked lists, binary trees, pointer arrays and other decisions (see e.g. [31] to illustrate the large search space which is available for this). In addition, virtual memory managers have to be defined to dynamically allocate or free these data in 'pools' within the memory space [30] (this step is also illustrated in [31]).

   This stage should be performed before the subsequent ones because in each of the next stages, several main steps depend on knowledge of the implementation cost for the major data types. And that cost cannot be accurately estimated or determined without a refinement of the initial abstract data types (either in the dynamic or static case). Also the virtual memory manager functionality has to be decided in order to get a reasonable idea of the allocation cost in the virtual memory space for the dynamic data types.

3. **task-level:** After the first stage, the designer ends up with an optimized system specification consisting of a number of concurrent communicating tasks, annotated with a number of timing requirements which define the boundaries for temporal correctness. Each of these tasks can contain one or more threads of control. Examples of this include the different coding tasks present in the MPEG-4 multi-media compression standard [26] oriented to audio, synthetic speech, video, synthetic images and other data streams. Also within e.g. the video coding part, several smaller tasks can be identified which act concurrently on the different data types present in the complex MPEG-4 algorithms (see e.g. video-object plane or VOP manipulation in figure 2.4). Due to the user interactivity and scalability (graceful degradation) which is explicitly supported in the MPEG-4 standard, many of these tasks also involve complex dynamic execution of tasks and data creation/destruction.

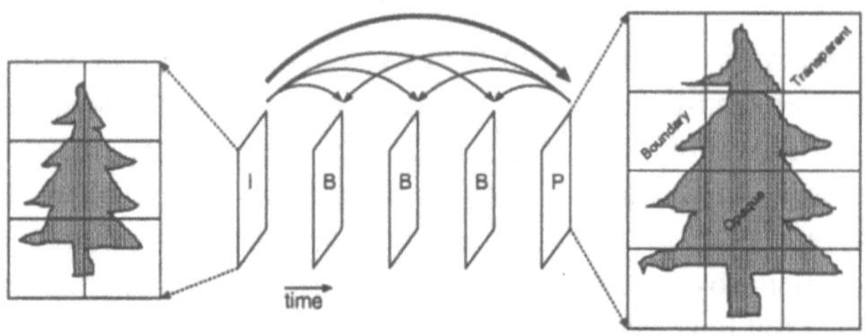

Figure 2.4 Illustration of MPEG-4 video object plane (VOP) sequence for content based object coding. The arrows represent all motion estimation steps for one group of VOPs.

First, the *task-level DTSE* stage should be executed where the DTSE steps (see above) are applied to the data selected at this task-level abstraction (see below). Examples of the importance of this stage for complex data-dominated applications are provided in [10, 11, 3]. A nice illustration of this stage is available in section 2.6.

The purpose of the subsequent *task concurrency management* stage is to determine an optimal constraint-driven scheduling, allocation and assignment of the various threads and tasks to one or more abstract processors, which on this level are called *array processors*.

4. **array level**: A task-level analysis only considers a limited amount of the potential parallelism of an application as the number of tasks and threads is usually rather small. In a practical design flow, many applications require that another level of concurrency is exploited, namely *data parallelism*. This is only true for applications which exhibit enough regularity (in some of their critical routines) and hence are based on arithmetic embedded in nested loop definitions. Moreover, the application should be run at a sufficiently large system data rate so that there is a need for parallelizing these nested loops over several "processor

elements" (PEs) by partitioning the data sets over distinct PEs. This does not imply massively parallel processing: only a few PEs are sufficient to motivate the inclusion of the array-level stages in the instantiated design flow.

A nice illustration of this is provided in chapter 5. Also the real-time execution of the video motion estimation kernel in a typical MPEG-4 coder context requires enough parallelism to motivate a weakly parallel array processor style. Still, in many other cases the array processor is simply a single heterogeneous processor and the instantiated design flow proceeds directly from task-level to instruction-level.

Also in this stage, we have to deal with data transfer and storage issues, so a *data-level DTSE* stage is foreseen. By extracting the possible data parallelism in a task/thread, the array processor can be further refined into a structure containing a number of *heterogeneous processors*, which each consist of one or more *homogeneous processors* (shown on the left of the instruction-level stage on figure 2.1). A homogeneous processor is characterized by only one *architectural style*, for example namely a type of custom or instruction-set processor (bit-serial, highly multiplexed, ...). In figure 2.1, the grid represents an array processor containing a number of these heterogeneous processors (the nodes in the array matrix). When the internal structure of the (physical) array processor has been decided, it is possible to partition the data and map the parallel data processing operations in the task/thread on this array processor. Work on this has been performed both in the systolic array synthesis community (see e.g. the chapters in [5]) and in the parallel compiler community (see e.g. [16] and its references). See also the material in chapter 5.

5. **operation/instruction-level**: Here, the operation/instruction-level concurrency issues are decided. Note that the custom processor synthesis and instruction-set processor mapping substages also synthesize the detailed interface descriptions (up to the RT-level) from the higher-level behaviour which has been specified during inter-task and inter-(homogeneous) processor interface refinement. Note also that this abstraction level is the minimal one which needs to be addressed for any realistic application. If no concurrent tasks are present and the application does not demand for an array-processor based realisation style, everything can be reduced to this stage in the instantiated meta flow.

   The latter situation is still true for most present-day applications in industry. Many applications can indeed be directly written as a single task, interacting with the environment, and array-processors are only needed for very demanding applications (for the time being). For a large part, this is however a decision of the system designer. Future complex applications are being (and will more and more be) specified as concurrent systems. The task-level stage is then vital to deal with the added design complexity and to fully exploit the design freedom which becomes available by adding this abstraction level. Note also that in principle, the concurrency extraction step at the task-level could be applied on an initial specification with a single task, leading to a concurrent (transformed) description for which the task concurrency management stage does become relevant. Likewise, all concurrent tasks which are present in a given specification could also be expanded into a single task if the designer decides to skip the task-level stages from the meta flow.

If the data types in a given application (domain) are not complex enough and hence are not dominating the impact of the concurrency issues, it is best to skip the DTSE stages at the higher abstraction layers (i.e. task and data concurrency DT SE steps) of the meta flow. Regarding the DTSE steps, the meta flow instantiation is then restricted to the processor-level DTSE stage for that application (domain). It will be assumed in the rest of this book that this is the case. If the cost (e.g. area or power cost function or weighted combination) associated with the most complex data types is not dominant (e.g. at least a factor 5 higher) over the (scalar) foreground signals, then even that processor-level DTSE stage should be skipped and all storage related issues have to be dealt with during the foreground memory management steps in the custom processor synthesis and/or instruction-set processor mapping stages. The latter are always needed, as any realistic application deals with at least some data type instances.

Issues on this level (see top of figure 2.5):

- **processor-level DTSE:** Storage and transfer exploration inside the scope of the heterogeneous processor. See section 2.6 for a nice illustration on the MPEG-4 video motion estimation kernel. Also chapter 3 mainly deals with this issue. A good overview of the state-of-the-art in this domain is provided in [7]. Apart from early work at IMEC [29], also Philips (see e.g. [18]) and Irvine (see e.g. [25]) have contributed to this area from quite early on.

- **processor architecture integration:** Here, the internal structure of the heterogeneous processor is decided in terms of homogeneous processors. An allocation of custom and instruction-set processors is made and the threads to be executed on this heterogeneous processor are scheduled on it directly based on the decisions made during the task concurrency management stage in terms of dynamic and static scheduling. Because all ordering decisions have been fixed earlier, the exploration of this aspect at this stage is reduced to the allocation/assignment only. Also the internal interfacing "glue" between the processors is generated during this stage. A good overview of the conventional way to look at this stage is provided in [15, 9].

  To illustrate this step in the MPEG-4 application, the video motion estimation kernel is e.g. best mapped on a set of custom processors. But the system layer steering it, should probably remain flexible enough to absorb modifications in the user requirements and context. So it should be mapped on a programmable instruction-set processor.

- **high-level address optimisation**: The mapping of multi-dimensional signals onto a one-dimensional memory structure is optimized. It precedes both instruction-set and custom processor mapping, since indexed signals can be handled in both of them. While it is true that most compilers generate these mappings automatically, they are rarely the best solution (e.g. a C compiler which translates an access of an n-dimensional signal by looking up a pointer in a table through n-way indirection). A much better way is usually to translate the n-dimensional index into a single address expression which is then aggressively optimized as an arithmetic expression before efficiently mapping it on the available arithmetic operators. An illustration of the search space and the important effects of this stage are provided in [22]. So high-level address optimisation lets the designer tune and optimize the array indices as a preprocessing step for the subsequent custom or instruction-set processor mapping stages.

In the MPEG-4 video motion estimation kernel, it is clearly crucial to have a cost-efficient realisation of all the necessary addresses, within the stringent real-time constraints. In chapter 3, an illustration is provided.

- **custom processor synthesis:** A custom processor is synthesized for the relevant parts of the heterogeneous processors. Reference books for this are available in e.g. [14, 4, 5].

  Chapters 5 and 7 of this book are largely dealing with this. Chapter 6 also addresses a dynamic power management strategy at this stage.

- **instruction-set processor mapping:** Instruction-set processor mapping (compilation and linking) is performed for the relevant parts of the heterogeneous processors. Optionally, the instruction-processor itself and its companion compiler are synthesized during this step.

6. **circuit level**: The circuit-level concurrency management stage maps the design on its *target technology platform*, which can e.g. be a reconfigurable (e.g. FPGA) target or a custom (e.g. ASIC) one. This stage is not a focus of this book.

A clear distinction is made between *target technology platform* and *target architectural style* (for example selecting a programmable or custom processor). These choices are made at two very different places in the meta flow. Abstraction is made of the terms "software" and "hardware" which have a connotation which does not correspond to an orthogonal choice at this stage [3]. Instead the terms *custom processor synthesis* and *instruction-set processor mapping* are used. This separation between the architecture level choice to either a custom or an instruction-set processor style and the technology platform selection (see figure 2.5) allows all 4 combinations to be used in the instantiated design flow. The conventionally used solutions namely exclude the combination of an instruction-set processor style mapped on a reconfigurable technology platform. Only recently this combination is beginning to be discussed and the meta flow clearly shows the feasible (sub)choices and the restrictions imposed by the dependencies between the substeps.

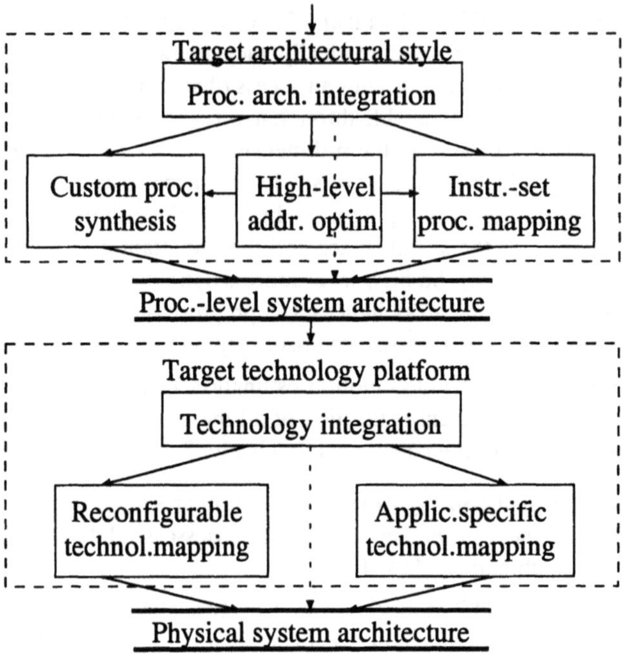

Figure 2.5 Architecture processor style decision versus technology platform selection.

2.6 ILLUSTRATION ON MPEG-4 VIDEO MOTION ESTIMATION DEMONSTRATOR

The separation of DTSE and concurrency issues, and the importance of working in different abstraction stages, are nicely illustrated in the exploration of the dominant video motion estimation module in the new multi-media compression standard MPEG-4. The recent MPEG-4 standard [26] is a key to multi-media applications. The exploration below has been performed on the MoMuSys Video VM Version 7.0. It involves complex data-dominant algorithms. An embedded instruction-set processor realization of such a (de)coder has to be power efficient in order to reduce the size of the chip packages (where it is embedded) or the battery (if used in a mobile application). We have shown that it is feasible to exploit the dominance of data transfer and storage related costs, to achieve large savings in the system power of a crucial part of MPEG-4, namely the video motion estimation (see figure 2.4), without sacrificing on the performance or on the system latency [2, 3].

Profiling experiments have shown that 77% of the original accesses take place for the luminance pixels in (full- and half-pel) motion estimation, which we have focused on[4]. The primary design goal has been to reduce memory transfers between large frame memories and processor data-paths.

We have focused only on the most promising code transformation steps in our methodology. In order to reduce the search space complexity, we have applied a two stage DTSE approach instead of our standard single stage. First, we have applied the DTSE transformations at the task-level abstraction where only a (relatively small) subset of the M-D signals has to be considered, which allows to reduce the design complexity in the global DTSE stage, and hence saving design time. By analyzing these task-level loop structures and large signals, the data locality of the algorithm implementation can be improved compared to the original sequence of motion estimation steps [26] (see also figure 2.4). However, not all motion estimation steps can be merged (infeasible due to data dependencies present in the actual code) and a thorough exploration is required of the huge search space, for which we have used our systematic loop transformation methodology.

This task-level transformed algorithm code result is now passed to the data reuse decision step [13]. In our systematic multi-step methodology several issues can be decoupled, allowing the potential reuse to be (more extensively) explored without letting the design complexity explode. At the task-level abstraction, only the main signals shared or communicated between the interacting tasks are considered. Moreover, only the first levels of data reuse are explored because the lower levels involve too small signal copies which have to be decided on during the more detailed processor-level DTSE stage. This leads to the results marked as task-level DTSE in figure 2.6. In this figure, the original specification is used as the reference point (with relative power 1). The left hand side steps belong to the task-level stage where only the main VOP signal access was considered. A reduction by a factor 65 in background luminance pixel reads has been obtained this way. Experiments have however confirmed that after this substantial saving on the VOP access, significant power is also related to the "second order" signals which are only visible when the abstraction level is expanded to the processor-level where the instruction/operation details become explicit again. Similar steps, applying our more conventional processor-level DTSE methodology (explained earlier), have been performed for each of the distinct tasks in the motion estimation code. We will not provide the details here. The right-hand side steps in figure 2.6 show that our processor-level DTSE code transformations allow to reduce also these contributions to a fraction of the original contribution. A substantial gain in global memory related average access power of more than a factor 8 is reached. This has been obtained and verified by simulation for the first 96 frames of the "hal" sequence in CIF resolution. The overall design time has been very reasonable due to our two-step approach.

The overall advantages of this new two-stage approach are that:

1. The task-level DTSE stage can have a truly global scope because only the most important inter-task signals (and only a few crucial intra-task ones) have to be considered. The processor-level DTSE stage can focus in full detail on each task separately without having to consider the task interaction any longer. These two facts result in significant savings on design time as experienced in our exploration of the MPEG-4 motion estimation routines.

2. In many applications, the concurrent tasks exhibit asynchronous process control interactions which prohibit the expansion of the full code into a single statically analyzable piece of code. In these cases, the detailed processor-level analysis and optimisations do not make sense beyond the boundaries of such tasks.

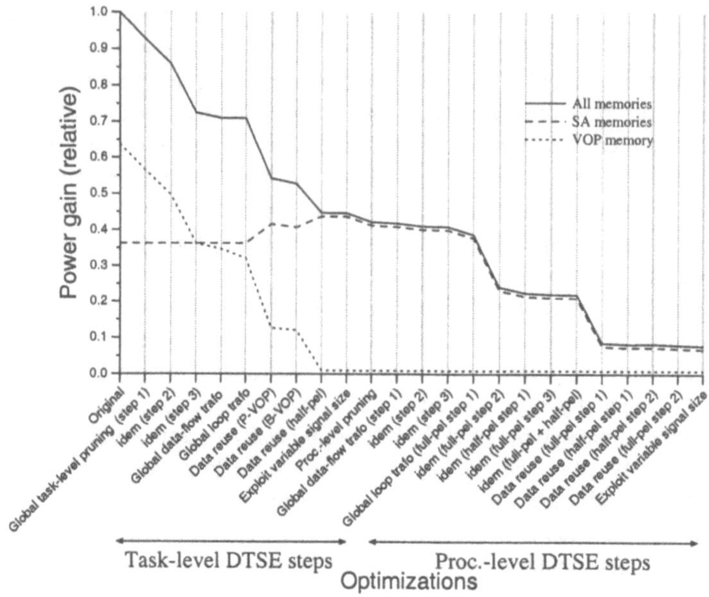

Figure 2.6 Total relative power for VOP memory and search area memories during application of task-level and processor-level DTSE stages. The power data (points) for the different steps in the optimisation stages have been connected by lines to easier see the main trends.

In contrast, our experiments have shown that the coarser view of the task-level DTSE stage still allows to perform relevant optimizations across the task boundaries.

The exploration space to optimize data storage and transfers is very large, and our experiments clearly show the importance of a formalized methodology to traverse it. It also demonstrates that the DTSE optimizations have to be performed aggressively before the multi-media algorithms are realized on custom or instruction-set processor targets. In general, we have experienced that it is well motivated to apply a DTSE stage prior to the concurrency management stage at any of the three abstraction levels in our unified meta flow.

2.7 PROPAGATION MODES THROUGH THE FLOW

Figure 2.7 shows the different propagation modes and paths through the flow. By ordering all the (sub)steps in the meta flow in a meaningful order, the information generated in them refines the system model at each level, while maintaining the right amount of abstraction, i.e. enough information to deal with the issues at that level without unnecessary details. This propagation of constraints through the meta flow links all the (sub)steps together in a coherent unified framework[5]. Because of the nature of this unified meta flow, no redundancy is really present between the objective of each (sub)step (up till the processor-level system architecture) so the combination of all of these constraints constitutes the final system architecture.

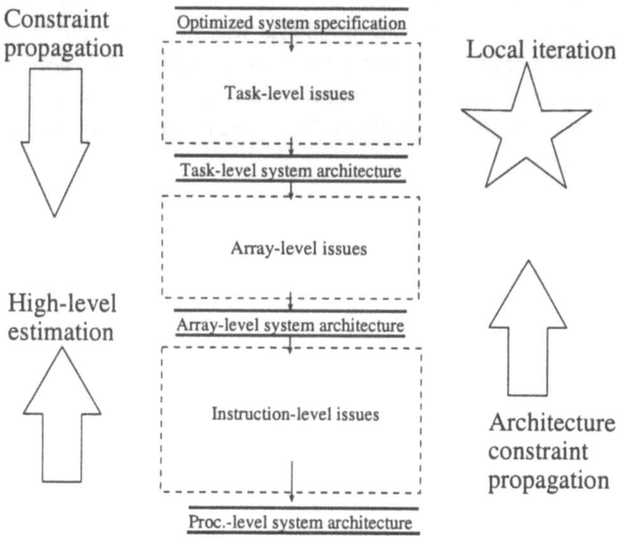

Figure 2.7 Iteration, constraint propagation and estimators

Note that besides the top-down constraint propagation mode outlined above, it is also likely that a designer has to deal with bottom-up architecture constraint propagation (see figure 2.7), for example when the decision to use a certain processor is made in advance. This will most likely lead to suboptimal results from a global cost viewpoint.

In each (sub)step, *high-level estimators* are typically used to predict what the main cost effect will be of the following (sub)steps. They guide local design iterations. Still, sometimes larger scope iterations will be necessary also, but these will be fully user driven and not triggered by information derived directly from the meta flow. However, the amount of iteration will be typically (heavily) reduced by applying the flow of the (sub)steps in the proposed order (as demonstrated by the many existing successful flows from which this unified one has been abstracted).

Note that in parallel to this unified system meta flow, there should also be a *system validation* (*system verification* and/or *system simulation*) flow and a *system-level test/debug* flow when errors are found. In principle, just as in the simulation/verification and software engineering community, these flows can closely follow the trajectory for the actual "synthesis". So in this book, it will be assumed that these other flows will follow the appropriate (sub)steps required to apply the instantiated meta flow in a given application.

2.8 CONCLUSIONS

In this chapter, we have outlined the relevant parts of our unified system design flow focused mainly on the low power aspects of different abstraction stages and illustrated by an application from the multi-media domain.

In the subsequent chapters of this book, most of these different stages will be treated in more detail, but the main emphasis will lie on the phases present in the instruction/operation-level stage, instantiated for a number of target application sub-domains, their corresponding architecture styles, and for the purpose of low-power design. We believe that the scope of validity of the complete material in this book, is much larger though.

Notes

1. A well-known example of this is the (operation-level) scheduling step. For real-time data-flow dominated applications, usually a list or force directed type scheduler is used. The list scheduling class is better matched to dealing with strict resource constraints, the force-directed techniques allow to better balance the data-flow in a given time budget when (almost) no resource constraints are present. Finally, for control-dominated applications, the path scheduling type provides a good option because it can deal well with conditional traces exhibiting largely varying critical paths.

2. No DTSE takes place at the circuit level, since there is no concept of "data"

3. Usually, one refers to an instruction set processor mapped onto an application-specific IC technology as a "software" processor

4. The cost of a data transfer is a function of the memory size, memory type, and especially the access frequency F_{real}. An accurate power model, derived from a VTI data sheet, has been used for the exploration.

5. These constraints are a key issue in the meta flow but the detailed list cannot be provided here.

References

[1] Multi-media compilation project (Acropolis) at IMEC `http://www.imec.be/vsdm/domains/designtechno/Welcome.html`

[2] E.Brockmeyer, F.Catthoor, J.Bormans, H.De Man, "Code transformations for reduced data transfer and storage in low power realisation of MPEG-4 full-pel motion estimation", *Proc. IEEE Int. Conf. on Image Proc.*, Chicago, IL, pp.III.985-989, Oct. 1998.

[3] E.Brockmeyer, L.Nachtergaele, F.Catthoor, J.Bormans, H.De Man, "Low power memory storage and transfer organization for the MPEG-4 full pel motion estimation on a multi media processor", *IEEE Trans. on Multi-Media*, Vol.1, No.2, pp.202-216, June 1999.

[4] R.Camposano, W.Wolf (eds.), "Trends in high-level synthesis", Kluwer Academic Publishers, Boston, 1991.

[5] F.Catthoor, L.Svensson (eds.), "Application-driven architecture synthesis", Kluwer Academic Publishers, Boston, 1993.

[6] F.Catthoor, F.Franssen, S.Wuytack, L.Nachtergaele, H.De Man, "Global communication and memory optimizing transformations for low power signal processing systems", *VLSI Signal Processing VII*, J.Rabaey, P.Chau, J.Eldon (eds.), IEEE Press, New York, pp.178-187, 1994.

[7] F.Catthoor, S.Wuytack, E.De Greef, F.Balasa, L.Nachtergaele, A.Vandecappelle, "Custom Memory Management Methodology – Exploration of Memory Organisation for Embedded Multimedia System Design", ISBN 0-7923-8288-9, Kluwer Acad. Publ., Boston, 1998.

[8] F.Catthoor, D.Verkest, E.Brockmeyer, "Proposal for unified system design meta flow in task-level and instruction-level design technology research for multimedia applications", *Proc. 11th ACM/IEEE Intnl. Symp. on System-Level Synthesis*, Hsinchu, Taiwan, pp.89-95, Dec. 1998.

[9] Special issue on "Hardware-Software Codesign", *Proc. of the IEEE*, Vol.85, No.3, March 1997.

[10] K.Danckaert, F.Catthoor, H.De Man, "System-level memory management for weakly parallel image processing", *Proc. EuroPar Conference*, Lyon, France, August 1996. "Lecture notes in computer science" series, Vol.1124, Springer Verlag, pp.217-225, 1996.

[11] K.Danckaert, F.Catthoor, H.De Man, "System level memory optimization for hardware-software co-design", *Proc. IEEE Intnl. Workshop on Hardware/Software Co-design*, Braunschweig, Germany, pp.55-59, March 1997.

[12] E.De Greef, F.Catthoor, H.De Man, "Program transformation strategies for reduced power and memory size in pseudo-regular multimedia applications", *IEEE Trans. on Circuits and Systems for Video Technology*, Vol.8, No.6, pp.719-733, Oct. 1998.

[13] J.P.Diguet, S.Wuytack, F.Catthoor, H.De Man, "Formalized methodology for data reuse exploration in hierarchical memory mappings", *Proc. IEEE Intnl. Symp. on Low Power Design*, Monterey, pp.30-35, Aug. 1997.

[14] D.Gajski, N.Dutt, A.Wu, "High-level synthesis: introduction to chip and system design", Kluwer Acad. Publ., Boston, 1992.

[15] D.Gajski, F.Vahid, S.Narayan, J.Gong, "Specification and design of embedded systems", Prentice Hall, Englewood Cliffs NJ, 1994.

[16] M.Hall, J.Anderson, S.Amarasinghe, B.Murphy, S.Liao, E.Bugnion, M.Lam, "Maximizing multiprocessor performance with the SUIF compiler", *IEEE Computer Magazine*, Vol.30, No.12, pp.84-89, Dec. 1996.

[17] K.Itoh, K.Sasaki, Y.Nakagome, "Trends in low-power RAM circuit technologies", special issue on "Low power design" of the *Proceedings of the IEEE*, Vol.83, No.4, pp.524-543, April 1995.

[18] P.Lippens, J.van Meerbergen, W.Verhaegh, A.van der Werf, "Allocation of multiport memories for hierarchical data streams", *Proc. IEEE Int. Conf. Comp. Aided Design*, Santa Clara CA, Nov. 1993.

[19] L.Nachtergaele, F.Catthoor, B.Kapoor, D.Moolenaar, S.Janssen, "Low power storage exploration for H.263 video decoder", *IEEE workshop on VLSI signal processing*, Monterey CA, Oct. 1996. Also in *VLSI Signal Processing IX*, W.Burleson, K.Konstantinides, T.Meng, (eds.), IEEE Press, New York, pp.116-125, 1996.

[20] M.C.McFarland, A.C.Parker, R.Camposano, "Tutorial on high-level synthesis", *Proc. 25th ACM/IEEE Design Automation Conf.*, San Francisco CA, pp.330-336, June 1988.

[21] M.C.McFarland, A.C.Parker, R.Camposano, "The high-level synthesis of digital systems", *Proc. of the IEEE*, special issue on "The future of computer-aided design", Vol.78, No.2, pp.301-318, Feb. 1990.

[22] M.Miranda, F.Catthoor, M.Janssen, H.De Man, "High-level Address Optimisation and Synthesis Techniques for Data-Transfer Intensive Applications", *IEEE Trans. on VLSI Systems*, Vol.6, No.4, pp.677-686, Dec. 1998.

[23] P.R.Panda, N.D.Dutt, A.Nicolau, " Memory data organization for improved cache performance in embedded processor applications", *Proc. 9th ACM/IEEE Intnl. Symp. on System-Level Synthesis*, La Jolla CA, pp.90-95, Nov. 1996.

[24] L.Rabiner, B.Gold, "Theory and Applications of Digital Signal Processing", Prentice Hall, Englewood Cliffs NJ, 1975.

[25] L.Ramachandran, D.Gajski, V.Chaiyakul, "An algorithm for array variable clustering", *Proc. 5th ACM/IEEE Europ. Design and Test Conf.*, Paris, France, pp.262-266, Feb. 1994.

[26] T.Sikora, "The MPEG-4 video standard verification model", *IEEE Trans. on Circuits and Systems for Video Technology*, Vol.7, No.1, pp.19–31, Feb. 1997.

[27] "Synthesis terminology catalog ", VSDM synthesis domains (eds. F.Catthoor, G.Goossens), Internal Documentation, IMEC, Heverlee, 1989 (first version).

[28] L.Terman, R-H.Yan (eds.), special issue on "Low power electronics" of the *Proceedings of the IEEE*, Vol.83, No.4, pp.495-700, April 1995.

[29] I.Verbauwhede, F.Catthoor, J.Vandewalle, H.De Man, "Background memory management for the synthesis of algebraic algorithms on multi-processor DSP chips", *Proc. VLSI'89, Int. Conf. on VLSI*, Munich, Germany, pp.209-218, Aug. 1989.

[30] P.R.Wilson, M.Johnstone, M.Neely, D.Boles, "Dynamic Storage Allocation: A Survey and Critical Review", *Proc. Intnl. Wsh. on Memory Management*, Kinross, Scotland, UK, Sep. 1995.

[31] S.Wuytack, J.L.da Silva, F.Catthoor, G.De Jong, C.Ykman, "Memory management for embedded network applications", *IEEE Trans. on Comp.-aided Design*, Vol.CAD-18, No.5, pp.533-544, May 1999.

3 LOW-POWER PROCESSOR-LEVEL DATA TRANSFER AND STORAGE EXPLORATION

Erik Brockmeyer, Cedric Ghez, Wim Baetens and Francky Catthoor

IMEC, Leuven, Belgium

brockmey@imec.be,ghez@imec.be,catthoor@imec.be

The current starting point of IMEC's systematic processor-level DTSE methodology [9, 10] is a system specification with accesses on multi-dimensional (M-D) signals which can be statically ordered. The output is a net-list of memories and address generators (see Fig. 3.1), combined with a transformed specification which is the input for the architecture (high-level) synthesis when custom realizations are envisioned, or for the software compilation stage in the case of predefined processors. The most time-consuming and error-prone steps within this methodology are becoming supported by tools, developed in the context of the ATOMIUM system exploration environment. The address generators are produced by a separate methodology called address optimisation or ADOPT (see section 3.12).

The research results on systematic exploration techniques and supporting prototype tools [1] which have been obtained in the last 8 years within the ATOMIUM project [9, 10] are discussed now.

3.1 MAIN PRINCIPLES

The DTSE methodology of IMEC comprises a gradual refinement of various issues. The steps are decoupled and placed in an order to reduce the number of iterations between the steps and to shorten the overall design time. The methodology minimizes in a systematic way the memory storage and transfers in order to reduce the power (and area) for data dominated applications in custom hardware architectures [48, 9]. To reduce the power consumption of an application we look again to the power formula ($P = C_{load} * f_{real} * V_{swing} * V_{dd}$). We can reduce both f_{real} and the effective C_{load} by the way we use the storage and interconnect devices. The goal should be to arrive

F. Catthoor (ed.),
Unified Low-power Design Flow for Data-dominated Multi-media and Telecom Applications, 25-63.

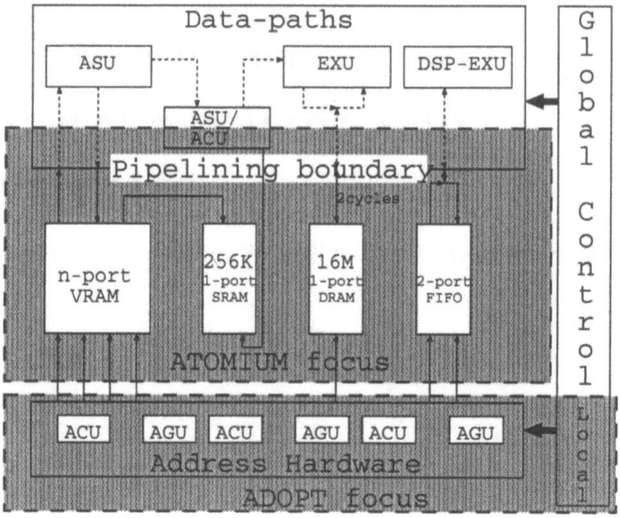

Figure 3.1 Current target architecture model for ATOMIUM and ADOPT: memory organisation and address hardware embedded in a global heterogeneous embedded system architecture.

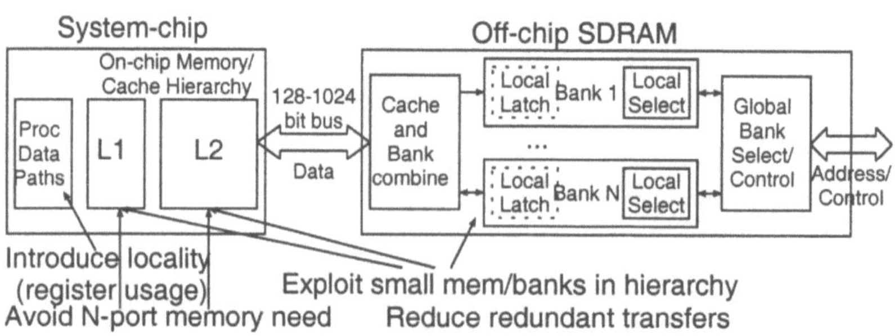

Figure 3.2 Main principles to reduce data transfer and storage bottle-necks in system-level data memory hierarchy.

at a memory and transfer organization with the following characteristics [14] (see also Fig. 3.2):

- Reduce the redundancy in data transfers (reduces f_{real})
- Introduce more locality and regularity in the accesses so that more data can be retained in registers local to the data-path (reduces f_{real} for the background memories)
- Use a hierarchical memory organization where the smaller memories (with reduced C_{load}) are accessed the most, and larger ones are accessed the least.
- Avoid the use of N-port memories if 1-port alternatives with a reasonable cost can be used instead (because of area, interconnect and power)

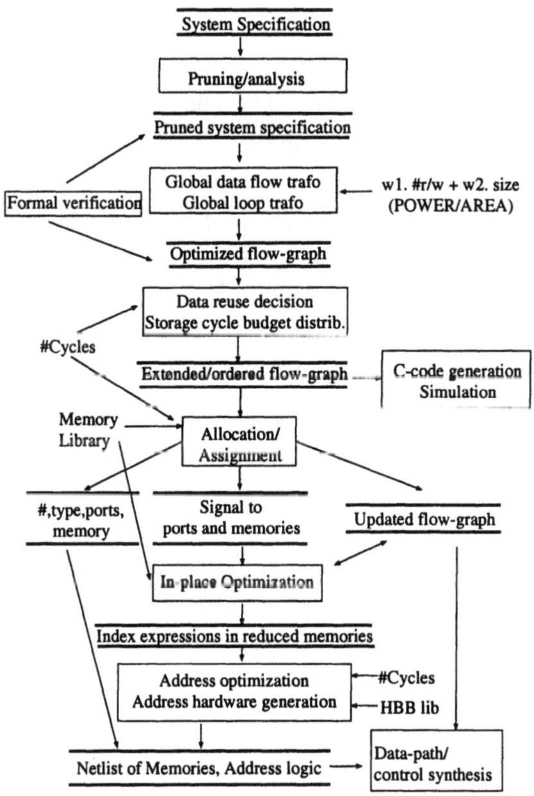

Figure 3.3 DTSE script for data transfer and storage exploration of the specification, to be used for simulation and hardware/software synthesis. This methodology is partly supported with prototype tools (in the ATOMIUM environment). The last task on address optimisation/generation is the target of a separate methodology and prototype tool-box (ADOPT).

These principles can be enabled by applying an appropriate system design methodology [3, 7, 8, 68]. The main steps involve (see also Figure 3.3):

1. To reduce the code to be optimized a preprocessing step is needed to focus on the important parts.

2. Reducing the required data bit-widths for storage by applying optimizing algorithmic transformations. These can change the actual algorithm choice, e.g. in digital filters or in the choice DFT versus FFT. Much can be done also in the step from floating-point data types to fixed-point types (see e.g. [8]).

3. Reducing transfer redundancy by data-flow transformations which do not modify the global input-output functionality [8]. This is a non trivial step because the redundancy is (mostly) not visible.

4. Increasing the regularity and locality by global loop transformations [18, 35]. Regularity will remove the buffers due to "dependency crosses". A global data access locality will create opportunities for more optimized data reuse.

5. Exploit memory hierarchy better by use of optimized decision on reuse of multi-dimensional data [13]. The data, from the main memory, will be copied to smaller buffers closer to the data-path. The reads from the smaller buffers are cheaper (in terms of power and speed) than reads from the (larger) main memory.

6. Balanced distribution of the globally available cycle budget over the loop nests such that the maximally required band-width is reduced [60]. This ensures that fewer multi-port memories are required and also fewer memories overall.

7. Allocate a more distributed memory organization [40] and assign the array signals cost-effectively to the different memories at each hierarchical level [60].

8. Exploit the restricted life-times of parts of the array signals to overlap them in the address space (in-place mapping) [22]. This enables a better exploitation of the caches and reduces the overall size of the required memories (and hence C_{load}).

More details are available in the cited references and the next sections. In the remainder of this chapter, they will all be illustrated on both educational examples and a real-life test-vehicle from the DAB-LP project of the European Low Power Initiative.

3.2 RELATED WORK

Up till now, little design automation development is available to help designers with this problem. Commercial EDA tools, such as SPW/Felix (Alta/Cadence Design), Monet/System Design Station (Mentor Graphics) and the COSSAP environment (Synopsys/CADIS), support system-level specification and simulation but are not geared towards design exploration and optimization of memory or communication-oriented designs. Indeed, all of these tools start from a procedural interpretation of the loops where the memory organisation is largely fixed, and they do not come back on these implicit design decisions. Moreover, the actual memory organisation has to be indicated by means of user directives or by a partial net-list. In the CASE area, represented by e.g. Statemate (I-Logix), Matrix-X (ISI) and Workbench (SES), these same issues are not addressed either.

In traditional (parallel) compiler research, memory related issues have been addressed in the past, but they have concentrated almost solely on performance in individual loop nests. They do not (yet) directly address energy and system bus load reduction, or more aggressive speed improvements obtainable by very global algorithm transformations.

In the parallelizing compiler community, work has focused on loop transformations to improve the locality in individual (regular) loop nests, e.g. at Cornell [39], at Illinois [5, 51], at Univ. of Versailles [27], and in the SUIF project at Stanford [2]. These do not work globally across the entire system, which is required to obtain the largest impact for multi-media algorithms, as we will show further on. Partitioning or blocking strategies for loops to optimize the use of caches have been studied in several flavours and contexts, in particular at HP [26] and at Toronto [36, 41]. Recently, also multi-level caches have been investigated [34]. The main focus of these memory related transformations has been on performance improvement in individual loop nests though and not on power or on global algorithms. As shown in Fig. 3.4, the speed depends on the critical path in the global condition tree, which can be viewed as a concatenation of local critical sections (bold line segments in the figure) in separate loop nests. For the power however, other weights are placed on the data-flow arrows and the global data-flow of complex data type accesses across conditional and concurrent paths has to be considered during the exploration for cost-effective code transformations. As a result, mostly different optimisation and exploration techniques have to be developed.

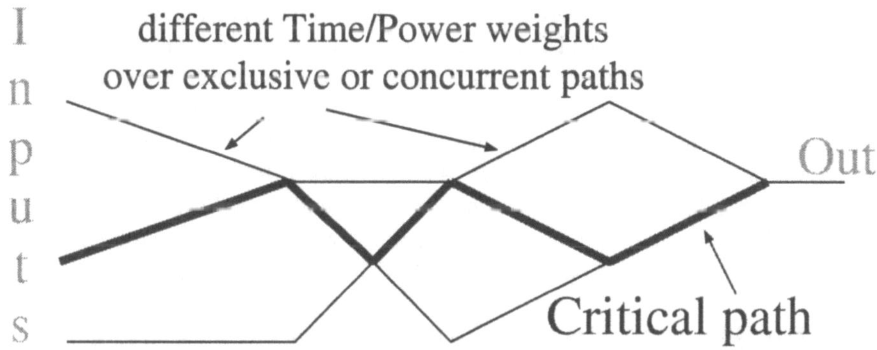

Figure 3.4 Illustration of difference between speed and power oriented optimisation. The speed exploration focusses on the critical path reduction, which operates on the worst-case path (line segments represent individual loop nests) in the global condition tree (with mutually exclusive branches). (Average) power optimisation should act on the global data flow over all concurrent and exclusive path.

In addition, as already mentioned our techniques work on a global scope in the search space available for the DTSE steps but this is only feasible by only focusing on the global data transfer and storage issues. So our approach is complementary to the traditional approach and it should be really seen as a source-to-source precompiler for the parallelizing or instruction-level compilers. Experiments further on will demonstrate that this decoupling is very well feasible and it leads to a nearly platform-independent precompilation stage in front of the conventional compilers [15] (see Fig. 3.5). Our approach also has a strong enabling effect on the use of subword parallel instructions in modern multi-media processors [42], so also for this purpose it is a crucial preprocessing step.

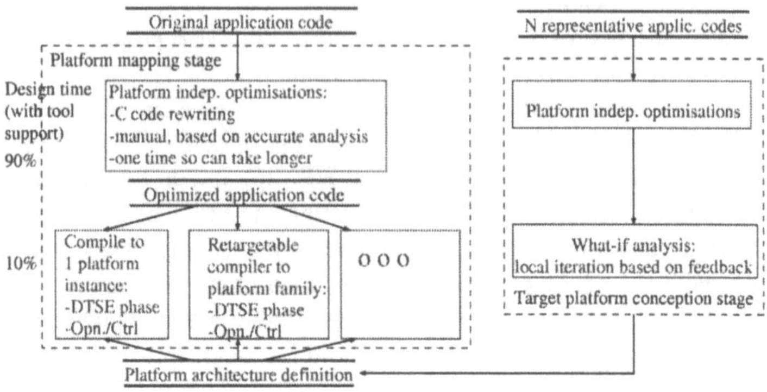

Figure 3.5 Systematic methodology for platform design with focus on DTSE aspects. Note the platform-independent steps in the initial stage and the tool support for the platform-dependent stage.

Within the system-level/high-level synthesis research community, the first results on memory management support in a hardware context have been obtained at Philips (Phideo environment [40]) and IMEC (prototype ATOMIUM environment [67, 48]). Phideo is mainly oriented to stream-based video applications and focuses on memory allocation and address generation. IMEC's ATOMIUM prototype design environment supports most memory related steps, either with interactive tools or in a more automated way, as discussed in this chapter. Also the work at Irvine, which started relatively early [53, 35], should be mentioned here. Recently they have started working on a compiler that is "memory-aware" [52]. Several recent initiatives in other research groups have been started up too (see e.g. [59, 58]), but they focus on point tools.

3.3 DIGITAL AUDIO BROADCAST APPLICATION

We have applied the DTSE approach to the synchronization processing of a Digital Audio Broadcast (DAB) demodulator and channel decoder. Also chapter 4 provides more results on the power distribution and the impact of the DTSE approach.

The DAB system is designed to provide reliable, multi service digital sound broadcasting for reception by mobile and stationary receivers, using a simple non-directional antenna. It is meant to replace the existing AM/FM networks, and to provide easy program selection and additional data services. The introduction to the public has occurred at the "Internationale Funk Ausstellung" in Berlin September 1995. The same month the BBC started a regular service containing all five radio programs in Great

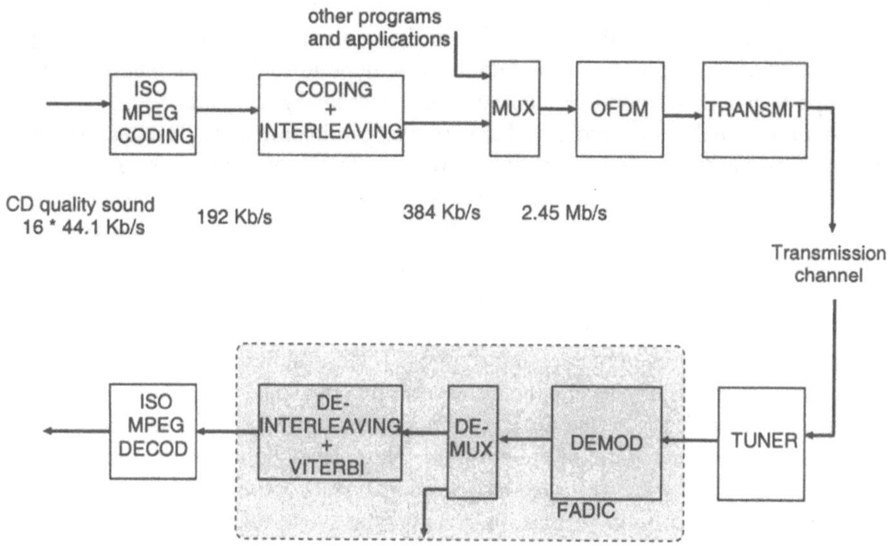

Figure 3.6 DAB system overview, the shaded area is the demodulator and channel decoder

London using a single frequency network. The single frequency mode is the first of the three operating modes in DAB, which is typically used for nation-wide coverage. The other two modes are meant for local and satellite broadcasting. DAB can be used from about 50 MHz until 2 GHz. The system is designed in the Eureka project 147 which has lead to an European Telecommunication Standard [10]. The DAB system is based on the technique of Coded Orthogonal Frequency Division Multiplex [11], in which demodulation is done using a Fast Fourier Transform. A more elaborate description of the DAB system can be found in [38] and [37] where experimental DAB receivers and integration issues are shown.

Figure 3.6 shows the complete DAB chain. Part of the shaded area is subject of this design case. This part (called synchronization processing) takes care of the synchronization between the receiver and the transmitter. It was designed using the architectural synthesis tool MISTRAL2.

3.4 MEMORY ORIENTED DATA FLOW ANALYSIS AND PRUNING

The background memory data can be partitioned into groups which are called Basic Groups (BG). In practice BG are declared arrays or parts of arrays and are treated in the same way by the algorithm (e.g. the odd and even lines of an image are 2 BG if handled separately). At the end, every BG data will be assigned as a whole to "virtual" memories. These are then later on mapped to physical memories but not necessarily in a one-to-one fashion. The main task of data flow analysis is to determine and characterize the BGs in the program and to potentially derive a maximally parallel flow graph at the BG level. The nodes in this flow graph represent the memory accesses to the data in the BG. This analysis has to be done at compile time and it provides the necessary granularity in the partitioning of data needed for independent ordering and assignment. This partitioning of data must be done in such a way that the resulting groups of data define non-overlapping ranges of memory accesses.

Accurate analysis is crucial to focus on the dominating part. The analysis has to happen on the full code and can therefore be a time consuming task. Atomium [3] is a tool which provides this analysis. Abstraction and analysis is provided to the designer with quantitative measurements on those characteristics of the application that are most critical in a multimedia context: memory accesses and data transfers. The analysis will be based both on classical profiling techniques and on state-of-the-art formal analysis techniques.

After a number of preprocessing steps [11], the pruning step divides the source into three levels. Every level contains several functions and procedures. Level 1 includes the main function, the different threads, the test-bench and support functions (e.g. loading of input data) of the program. Next, level 2 contains the indexed array signals augmented with the nested loops and manifest condition scopes in which they are produced and consumed. Last, level 3 contains the lower level functions like the: data path related issues, arithmetic functions and logical operations on scalars or small arrays (including the local control). Only, the level 2 code is treated by the DTSE. During this exploration, the level 3 functions are untouched, only the in/output dependencies are important. Therefore, the actual level 3 function content is irrelevant. In order to use data-flow analysis, the important signals have to be "single assignment". This means that every (scalar element of a) signal is defined exactly once in each conditional control scope. Similarly, "single definition" is less strict: every (scalar element of a) signal is defined at most once. For instance while construction has not to process all elements of an array. Here, the more relaxed "single definition" is adopted to avoid non necessary overhead.

So why pruning, and why is a large pruning factor needed? First, the focus should be on important code, and by performing this pruning step you achieve a better global view on the whole core. Secondly, the Polyhedral Dependence Graph (PDG) [66, 21] based techniques which are embedded in several of our supporting tools, can only handle a few pages of code within interactive run-times. This means that *either* the initial code has to be heavily pruned *or* it has to be partitioned in "functions" where the subdivision in functions is procedurally interpreted. The latter option directly leads to a local view, i.e. all calls to a given function use the same HLMM based organisation for the content of the function. This option is inherently easier to support but it leads to a large reduction in the potential search space as all the mismatches between production and consumption of data are then fixed, leading to buffers in between the functions executions later on. Practice has shown that the biggest gains are situated in these buffers and they can only be optimized by a very global view. This means that code up to several dozen pages pruned code should be dealt globally with for advanced applications. Thirdly, nonlinear and non-optimizable data dependent local control can be hidden in the level 3 code that the DTSE steps and tools do not have to deal with them. Finally, this explicit subdivision eases verification [13] and even design reuse [65].

It should be noted that the term "pruning" should be used with care because it is a common name for several tasks in the full script. Some of these are better classified as "preprocessing" but they have been clustered here under the term pruning.

The remainder of this section illustrates the pruning subscript on the IFFT module in the DAB synchro core. First, a pruning example in the C language is given. The DTSE methodology is not restricted to the C language, but this is the de facto standard which is used by most application designers who model executable code. Secondly, we will demonstrate how the C code is obtained from another language while preserving the bit-true behaviour. It shows that the translation of the code can be a limited part of the complete specification. Indeed, this limitation saves significant design time.

An example of unpruned and corresponding pruned C code is given in Figure 3.7. The left figure shows the original IFFT code from the synchro core. The data-path related issue, here the butterfly calculation, is moved to a level 3 function (see right-bottom Figure 3.7). The total code size can locally increase somewhat due to the additional level-3 functions, but experience indicates that this remains very acceptable for the complete application [43]. However, the relevant level 2 part will decrease in size. Remember, that DTSE does not change any data-path related code. Therefore, pruning has reduced the amount of code to optimize. Another goal of pruning is improving readability and transformation ease. To this end the indices will be written in an explicit way and pointers are avoided. The iterative address calculation is a burden during the transformation stages. Although the address calculation is very efficient it has been made explicit. As a result, the addressing part of right Figure 3.7 now only contains constants and loop iterators. Finally, the loops, array accesses and address calculation are the only remaining constructs. Hence, the relevant DTSE-oriented code size for the entire synchro core is reduced by a factor 4.

A wrapper around the DTSE is needed to import specifications from other languages. The pruning is useful to minimize the amount of code translated in both directions. The original specification of the synchro core is given in the Data-Flow Language (DFL) of Frontier Design. In the first step (see Figure 3.8), all data-path related issues (adding, shifting, multiplication, bit manipulations etc.) are hidden in level 3 functions. The bit-true behaviour is preserved by inserting function calls for the data-path operations. Secondly, the level 2 DFL is translated to C. A Perl script has been written to support this task. Similarly, the level 3 code is translated. However, strictly this is not needed since dummy functions could be used instead. In addition a test-bench is written which feeds data into the synchro core function. After the DTSE optimizations, the optimized level 2 code has to be translated back (to DFL). Since the level-3 code is unchanged, only the level-2 code has to be translated, saving time again. The level-3 code can easily be inserted to complete all functionality. Finally, function inlining, fine tuning and other local optimisations are needed to improve the core performance.

3.5 GLOBAL DATA-FLOW TRANSFORMATIONS

Two classes of data-flow transformations exist. The first class aims at removing redundant data storage and accesses. Four different subclasses of data-flow transformations of this class have been identified and they are best handled in the following order[2]:

Unpruned C source

```
/* stage loop */
for (stage=0; stage<n_stage2_ifft; stage++)
{
  addr0_ifft = -1;
  k_ifft     = -1;
  /* block loop */
  for (block=0; block<n_block_ifft; block++)
  {
    k_ifft  = acuadd(k_ifft,1);
    /* calculate the 2 sample addresses */
    addr0_ifft = acuadd(addr0_ifft,1);
    addr1_ifft = acuadd(addr0_ifft,n_butterfly_ifft);
    /* get the twiddle factor */
    twiddlecos_ifft = cos[k_ifft];
    twiddlesin_ifft = sin[k_ifft];
    for (butt=0; butt<n_butterfly_ifft; butt++)
    {
      /* radix-2 butterfly: */

      /* read data */
      br_ifft = memr[addr1_ifft];
      bi_ifft = memi[addr1_ifft];

      /* perform complex muliplication */
      mr_ifft = times(br_ifft,twiddlecos_ifft);
      mr_ifft = macn(mr_ifft,bi_ifft,twiddlesin_ifft);
      mi_ifft = times(bi_ifft,twiddlecos_ifft);
      mi_ifft = macp(mi_ifft,br_ifft,twiddlesin_ifft);

      /* for rounding of cr,ci,dr,di : add always +1 */
      /* (is only incorrect for -xxx.5000 )       */
      ar_ifft = (memr[addr0_ifft]) + 1) << 8;
      ai_ifft = (memi[addr0_ifft]) + 1) << 8;

      /* perform addition and subtraction */
      cr_ifft = (ar_ifft + mr_ifft) >> 9;
      ci_ifft = (ai_ifft + mi_ifft) >> 9;
      dr_ifft = (ar_ifft - mr_ifft) >> 9;
      di_ifft = (ai_ifft - mi_ifft) >> 9;

      /* write data: */
      memr[addr0_ifft] = cr_ifft;
      memi[addr0_ifft] = ci_ifft;
      memr[addr1_ifft] = dr_ifft;
      memi[addr1_ifft] = di_ifft;

      /* setup for next iteration: */
      addr0_ifft       = acuadd(addr0_ifft,1);
      addr1_ifft       = acuadd(addr0_ifft,n_butterfly_ifft);
    } /* butt */
    addr0_ifft = acusub(addr1_ifft,1) ;
  } /* block */
  n_butterfly_ifft = downsh(n_butterfly_ifft);
  n_block_ifft     = acuadd(n_block_ifft,n_block_ifft);
} /* stage */
```

Pruned C source

<level 2>, loops, accesses, adresses, dependencies

```
for (int stage = 1; stage <= n_stage_ifft; stage++)
{
  for (int block = 1; block <= ( 1<<(stage-1)); block++)
  {
    for (int butt = 1; butt <= (size_ifft>>(stage-1)); butt++)
    {
      /* calculate addresses */
      addr0_ifft = ((block-1)*((size_ifft>>(stage-2)))+(butt-1));
      addr1_ifft = ((size_ifft>>(stage-1))+
                   ((block-1)*((size_ifft>>(stage-2))))+(butt-1));
      k_ifft     = (block-1)*(size_ifft>>(stage-1)) + butt;

      /* perform butterfly */
      Radix2ButterFly(memr[addr0_ifft][stage-1], memi[addr0_ifft][stage-1],
                      memr[addr1_ifft][stage-1], memi[addr1_ifft][stage-1],
                      &memr[addr0_ifft][stage], &memi[addr0_ifft][stage],
                      &memr[addr1_ifft][stage], &memi[addr1_ifft][stage],
                      cos[k_ifft],              sin[k_ifft]);
    } /* butt */
  } /* block */
} /* stage */
```

<level 3>, data-path, bit-true related issues

```
void
Radix2ButterFly(int  ar_ifft, int  ai_ifft,
                int  br_ifft, int  bi_ifft,
                int& cr_ifft, int& ci_ifft,
                int& dr_ifft, int& di_ifft,
                int twiddlecos, int twiddlesin)
{
  /* perform complex muliplication */
  mr_ifft = times(br_ifft,twiddlecos_ifft);
  mr_ifft = macn(mr_ifft,bi_ifft,twiddlesin_ifft);
  mi_ifft = times(bi_ifft,twiddlecos_ifft);
  mi_ifft = macp(mi_ifft,br_ifft,twiddlesin_ifft);

  /* for rounding of cr,ci,dr,di : add always +1 */
  /* (is only incorrect for -xxx.5000 )       */
  ar_ifft = (ar_ifft]) + 1) << 8;
  ai_ifft = (ar_ifft]) + 1) << 8;

  /* perform addition and subtraction */
  cr_ifft = (ar_ifft + mr_ifft) >> 9;
  ci_ifft = (ai_ifft + mi_ifft) >> 9;
  dr_ifft = (ar_ifft - mr_ifft) >> 9;
  di_ifft = (ai_ifft - mi_ifft) >> 9;
}
```

Figure 3.7 Pruning example of an FFT. The pruned level-2 code contains only loops, memory accesses and explicit address computation. The data-path operations are hidden in level-3.

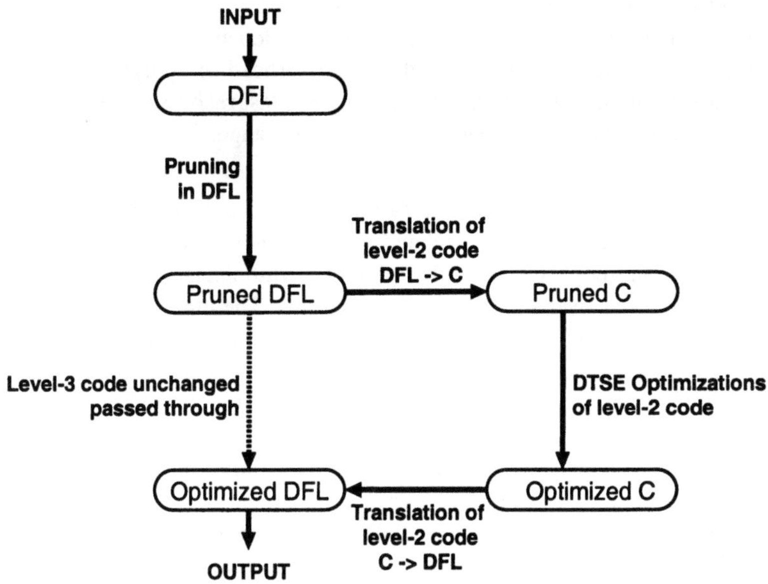

Figure 3.8 Preprocessing, translations from DFL to C and back

1. **Advanced re-substitution including moving conditional scopes.**
   Use re-substitution to remove dependencies on signals which are "copies" of other signals and which would normally have to be separately stored. This option is already applied during pruning but in general this transformations can have a much more complex global effect.

2. **Modifying computation order in associative chains**
   The computation order in associative chains like accumulation can be transformed to have more locality of access. An other example of using these associative chains is to decrease the word-length of the intermediate data storage at the places in the flow graph where the most data "pass". These transformations can effect the bit-true behaviour so in general the conditions under which this can happen have to checked. If done carefully, this can however be guaranteed to work correctly without re-simulations.

3. **Shifting of "delay lines" through modules.**
   E.g. to move a delay to an edge with a smaller word-length or to an edge with less signal words or to move several delays trough the graph by re-timing, generalized to handle delays on array signals, to reduce the total associated storage.

4. **Re-computation issues.**
   E.g. recompute a large amount of data from a limited amount of "key data", if the large amount of data is used multiple times. This leads to less global transfers, but is only useful in a context where long life-times would not lead to too much extra storage. It involves a power-area trade-off. This stage of transfor-

mations is very sensitive to the estimated storage and access cost at different positions in the algorithm. Hence, it is very dependent on the outcome of all the previous stages. Moreover, the area overhead of applying it is in general relatively high, so it should only be used as a last recourse to reduce the power cost. Hence, it should be performed as the last stage.

The correlator in the DAB synchronisation core is nice illustrative example for this first class. The input symbol is correlated with the known synchronisation reference symbol to find the frequency error. The reference symbol has a length of 16 elements and the correlation is over a range of 127 positions[3]. The input array is preprocessed (complex multiplication) before before correlated. A down scaled correlator (length of correlation reduced to four) is shown in Figure 3.9. The preprocessing is recalculated for every element needed in the correlation. Importantly, two input array elements are needed to preprocess one basic correlation. On the other hand, the preprocessed elements can be stored. Besides the initial overhead, only a single read is needed for every basic correlation. In the figure, every input element is reread eight times for the recalculation case (left) and the intermediate storage implementation case seven times (right). In the full scale implementation these access counts are respectively 32 and 19. Thus a gain of 40% is obtained at the cost of a larger chip area (caused by the intermediate storage array).

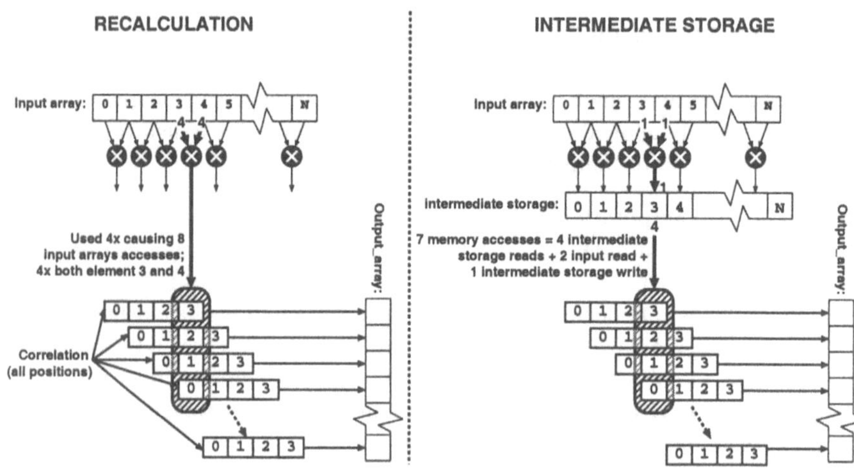

Figure 3.9 Illustration of a recalculation data-flow transformation on a scaled DAB correlation example. In case of recalculation (left), every input array element is read eight times. When storing intermediate data (right) the number of accesses is reduced to seven. In the full DAB case a gain of 40% is obtained.

The second class allows to remove data bottle-necks that hinder global (loop) code movement. This is a crucial enabling step for the rest of the DTSE script, which can be partly merged with the subsequent loop transformation step. In the latter, the production and consumption of array signals are put as close as possible to each other.

This will allow to decrease the required buffer size or even completely remove the buffer between the producer and the consumer of data.

For instance in the DAB synchro core context, the maximum calculation of the weight channel impulse response and the last stage of the IFFT calculation cannot be merged before breaking the iterative weight calculation. Doing so will prevent the large buffers between these functions and the accesses to these buffers. Due to the IFFT data dependencies, the channel impulse response is produced in a address bit-reversed order. This is overcome with the "bit-reverse address shuffle" module before the actual maximum calculation (see Figure 3.10). This introduces storage and access overhead. To avoid this, this intermediate shuffling module can be merged into the weighted maximum calculation step by substituting the index calculations. In addition, the overall access locality improves further when the last stage of the IFFT is merged with the maximum calculation module. However, the weight is calculated in an incremental way to enforce a linear order. Fortunately, neither the last stage of the IFFT nor the maximum calculation module, enforce any specific execution order so they can be adapted to the required linear order. Indeed, the maximum function is associative and the FFT butterflies of the last stage can be performed in any order. Still, a problem remains because the butterfly produces a pair of output elements which must both be used by the maximum calculation for optimal locality. Due to the bit reversal, this butterfly element pair is separated over half the size of the IFFT addresses (see Figure 3.10). Therefore, only one element can be used immediately and the other needs to be stored temporarily, except when the other weight factor is calculated as well. Indeed, two separate weight values coexists and are both calculated in an incremental way. The initial value of the second weight can be pre-calculated and substitutes the first half of the incremental calculation (data-flow transformation type 1, see Page 33). Actually, the additional weight calculation improves the ordering freedom because it introduces more concurrency in the dataflow. The end result is a removal of the data-flow bottleneck so the subsequent global control flow optimisation step can heavily improve the access regularity and locality.

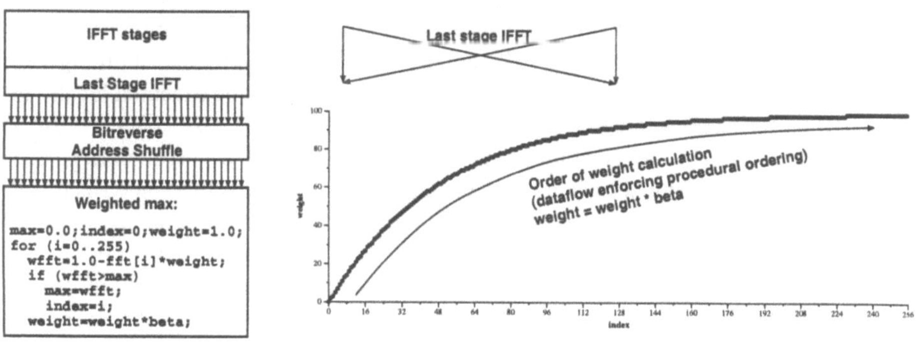

Figure 3.10 Illustration of a bottleneck breaking data-flow transformation on the DAB synchro demonstrator. The weight calculation initially enforces a procedural ordering. This ordering (bottleneck) can be broken by substituting the weight.

3.6 GLOBAL LOOP AND CONTROL FLOW TRANSFORMATIONS

Both loop transformations (unimodular ones, more general affine ones and even non-affine ones) and other reindexing transformations modify the index structure of array signals. The target of the loop transformations is to bring the production of data as close as possible to (all) their consumption(s). In contrast to the data-flow class, the loop transformations modify only the index structure, changing the order of execution. They do not effect the bit-true behaviour, so no re-simulation is needed afterwards when applied correctly. The latter can be checked with formal verification techniques [13].

In the past, an interactive loop transformation prototype engine (SYNGUIDE) has been developed by us that allows both interactive and automated (script based) steering of language-coupled source code transformations [54]. It includes a syntax-based check which captures most simple specification errors, and a user-friendly graphical interface. The transformations are applied by identifying a piece of code and by entering the appropriate parameters for a selected transformation. The main emphasis lies on loop manipulations including both affine (loop interchange, reversal and skewing) and non-affine (e.g. loop splitting and merging) cases.
In addition, research has been performed on loop transformation steering methodologies for an embedded custom or instruction-set processor context where not only performance is crucial. For power, we have developed a script oriented to removing the global buffers which are typically present between subsystems and on creating more data locality [19, 12]. In principle, this can be applied manually. In the ACROPOLIS context, an automatable compiler technique is currently being developed [16].

Regularity and locality are important cost functions at this level of optimization. A poor locality causes too large buffers in the final implementation. A bad regularity will enlarge the numbers of buffers (in the form of double buffers) and impose a complex design. The good regularity of the traditional FFT implementation (see left Figure 3.11) allows that all the stages can be in-placed and can make use of a single buffer. Unfortunately, the bad locality prohibits the usage of smaller buffers in the first FFT stages (see also Subsection 3.11). The locality of the FFT can be improved [4] by splitting the FFT in two (or more) FFTs and by introducing an irregular shuffling step in between (see right Figure 3.11). The smaller FFTs can now be executed locally in a small buffer or even foreground memory (here, the local FFT size is eight). But a major disadvantage is now that the irregular shuffle will introduce a second large and costly buffer. Therefore, these kind of dependency crosses should be avoided as much as possible or moved to "cheap" locations in the algorithm. In our final algorithm, a good trade-off between these 2 extremes is found (see Subsection 3.11).

The DAB correlator can be loop transformed in such a way that the cost bottleneck is moved from one place to another. The target is to minimize the global cost. In the original implementation, the largest portion of array accesses are to the *input_array* (see left Figure 3.12). For every output element, four input elements are accessed. By interchanging the two inner loops, the *output_array* is accessed four times per single input array element. The high access count should be moved to the smallest array to obtain the lowest power. The cost of an array access is dependent on the array size and

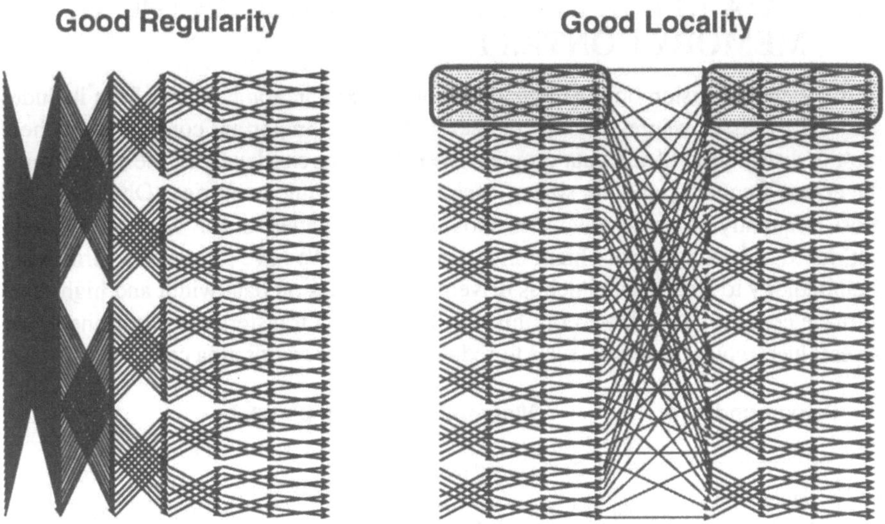

Figure 3.11 Regularity and locality are important costs for optimizing the index structure.

bit-width. The *input_array* is much larger than the *output_array* but the bit-width of the latter is much higher. After consulting an accurate memory model (depending on the target process technology), a gain of 26% is obtained by putting most accesses to the *output_array*.

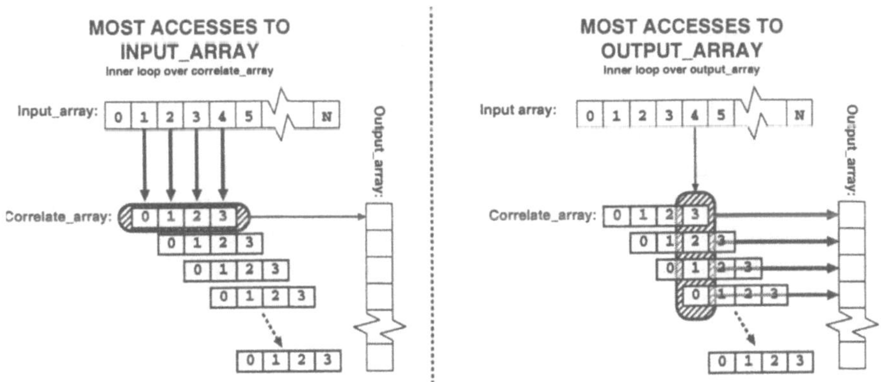

Figure 3.12 A loop interchange in the correlator exchanges the *input_array* memory accesses for *output_array* memory accesses. This trade-off can be optimized using accurate cost models.

3.7 DATA REUSE DECISION IN A HIERARCHICAL MEMORY CONTEXT

In this step, a decision on the memory hierarchy exploitation is made. This includes also the bypasses wherever they are useful [69, 71]. Important considerations here are the distribution of the data (copies) over the hierarchy levels as these determine the access frequency and the size of each of the resulting memories. Obviously, the most frequently accessed memories should be the smallest ones. This can be fully optimized only if a memory hierarchy is introduced. We have proposed a formalized methodology to steer this, which is driven by estimates on bandwidth and high-level in-place cost [13, 71]. Based on this, the background transfers are partitioned over several hierarchical memory levels to reduce the power and/or area cost.

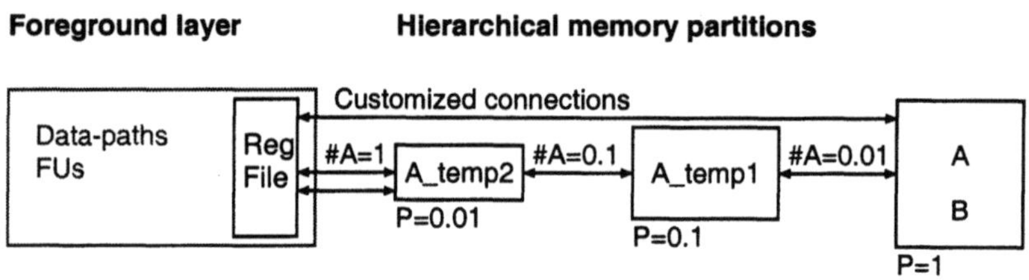

Figure 3.13 Memory hierarchy illustration: the foreground memory next to the data-paths consists of registers and register files; the intermediate buffers (partition $P1$ and $P2$) typically consist of fast embedded synchronous SRAMs with many ports to the lower level where potential "signal copies" from the higher levels are stored; the top layer (partition $P3$ in this example) consists of a slower mass storage unit with low access bandwidth.

A small illustration of this is shown in Fig. 3.13, where two signals, A and B, are stored in background memory. Signal A is accessed via two intermediate signal copies (A_temp1 and A_temp2), so its organisation involves three levels, while signal B is accessed directly from its storage location, i.e. located at level 1. Notice that A and B are stored in the same memory. This illustrates that a certain memory can store signals at a different level in the hierarchy. Therefore we group the memories in partitions ($P1$, $P2$, and $P3$) instead of "levels". The data reuse and memory hierarchy decision step selects how many partitions are needed, in which partition each of the signals (most of which are only intermediate data in RMSP applications) and its temporary copies will be stored, and the connections between the different memory partitions.

An important task in this step is to perform transformations which introduce extra transfers between the different memory partitions and which are mainly reducing the power cost. In particular, these involve adding temporary values – to be assigned to a "lower level"– wherever a signal in a "higher level" is read more than once. This involves clearly a trade-off between the power (and partly also memory size) lost by adding these transfers, and the power gained by having less frequent access to the larger more remote memories at the higher level(s).

The potential energy gain is large, while introducing a small overhead in memory area. For instance, the energy gain for signal A is a factor 30 in the example above. The signals A_temp1 and A_temp2 are much smaller and can be stored in smaller memories which in this illustration are assumed to consume respectively 10% and 1% of energy per memory access compared to the main background memory (denoted by the $P =$ .. under the box). These are quite representative numbers because off-chip memories easily consume an order or magnitude more power then on-chip memories. For low-power reasons, most memory accesses take place on the smaller buffers. Ofcourse, the number of data elements transferred to the data-path remains equal (denoted with #A=1). In the example, enough data reuse is present so that only 10% of the data elements need to be retrieved from the second level (#A=0.1) and 1% from the main memory (#A=0.01), after the data reuse transformations. Without this memory hierarchy, all the memory accesses are retrieved from the main memory dissipating a relative energy of 1.0. When including the proposed hierarchy, the relative dissipated energy is only 0.03 (the main memory→0.01*1.0, A_temp1 →0.1*0.1, A_temp2 →1.0*0.01), illustrating the large potential savings.

In order to investigate every possible solution for memory hierarchy, we have introduced data-reuse trees. How do we have to make such a reuse trees? Starting from the loop-bounds as size of your intermediate signal copy, incorporating more loops from a nest introduces more levels of signal copies. You can construct a reuse tree with all these signal copies. This is actually a compact overview of all possible solutions for a memory hierarchy. Every possible useful hierarchy is embedded somewhere in these tree. Many inherently suboptimal solutions can be pruned already during the construction of these trees because for every extra loop and corresponding intermediate signal copy that are considered, it is checked whether enough data reuse is available to compensate for the additional transfers for copying. When the cost for a read and a write on every level is known, it is possible to calculate which global signal copy chain is the cheapest for power reasons. It is also possible now to make a trade-off with other costs like the area consumption. More about these reuse trees can be found in [13]. After the pruning of these data reuse trees, they have to be combined for all signals that are later assigned to a common memory hierarchy. Finally, the actual memory hierarchy organisation has to be decided. This last step is already target platform and memory library dependent.

Again, the correlator in the DAB synchronisation core is used as an illustrative example. Remember, the reference symbol has a length of 16 elements and the correlation is over a range of 127 positions. Therefore the number of memory accesses to the data-path is 127×16=2032 per correlation; normalized as 1.0 in Figure 3.14. Also the power consumed by and area needed for this memory is normalized as 1.0. When the correlator moves to the next position 15 of the 16 elements are equa and can be reused in a smaller memory of 16 elements. As a result the number of memory accesses is nearly reduced with a factor 16 (7% of original power). Most accesses can take place on this smaller memory and consume less power. However, additional memory writes are introduced and bus overhead. Taking this into account, the intermediate memory will consume 80% of the original full size memory. Summing the power will reduce the total power with 13%. As a cost, the additional intermediate memory increases the area with 9%. Of course these figures are very dependent on the memory library used.

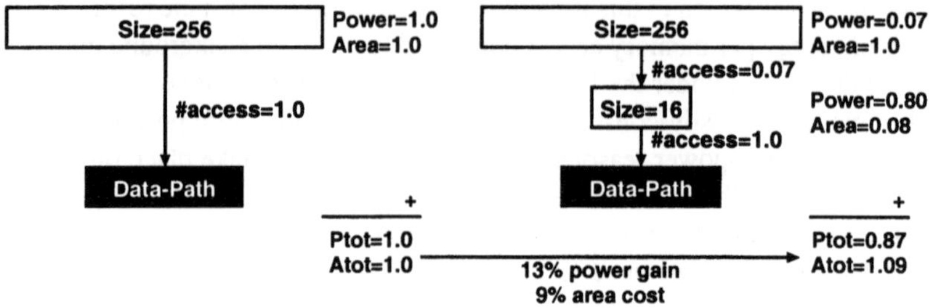

Figure 3.14 Data-reuse example in the context of the DAB synchronisation core. When inserting an intermediate memory 13% power can be saved at the cost of 9% area increase.

3.8 MEMORY BANDWIDTH OPTIMIZATION

When the memory hierarchy has been chosen, the physical memory management stage itself can begin. Physical memory management involves the design of a custom memory organisation for an application. It is clearly platform-dependent. The memory organisation must be as cost-efficient as possible, but on the other hand the performance must be just high enough to meet the real-time constraints.

The first step in physical memory management is to make the trade-off between the cost of providing a certain amount of memory bandwidth, and the real-time constraints. The approach taken in storage cycle budget distribution is to order the memory accesses such that the maximally required memory bandwidth is minimized. This can be done by (more) efficient use of the needed memory ports. In our application domain, an overall target storage cycle budget is put forward, corresponding to the real-time constraints.

The result of the storage cycle budget distribution step is an accurate estimate of the required global memory bandwidth. When two basic groups are accessed in parallel (Figure 3.15), an edge between those two basic groups is added in the conflict graph. This means that the two basic groups cannot be placed in the same memory — if it is a single-port memory, at least.

To take into account multi-port memories, the conflict graph has to be extended. In this case also self-conflicts are important namely, because they force to use multi-port memories (see Figure 3.15 for basic group $memr$). A second extension is the annotation of the number of parallel accesses. These are the number of parallel accesses (#read/#write#read-write) and are placed at the edges. Hyper edges show conflicts between multiple basic groups. For instance, when three basic groups are accessed in parallel then they are all in conflict with each other. However, according to the conflict graph the three signals can be put in a dual-ported memory. The combined conflict (hyperedge) prevents this illegal binding. In Figure 3.15 a hyper edge is represented as a square box and is connecting e.g. $memi$, $memr$ and $tfprref$. This implementation is efficiently represented in our extended conflict graph for memory accesses [70, 60, 9].

We have developed the storage cycle budget distribution step to formalize the trade-off between the time assigned to a given loop or task, and the cost (in this book mainly energy or power) that is required for that part of the global application. For flat flow-graphs the technique has been published in [70, 60] and [9, chapter 10], and it is known as *flow-graph balancing* (FGB). More recently, it has been extended to deal with multiple basic blocks. Basic blocks are parts of the code which are a single flat graph (no loops, no functions, no hierarchy). The experiments in this chapter rely on preliminary work in this area, implemented within a prototype tool to automate part of the work to be done.

This cycle budget must be distributed over the different loop nests, resulting in a cycle budget for every basic block. This is needed in addition to ordering the memory accesses within the basic blocks (using the earlier mentioned FGB technique). Moreover, it is beneficial for the overall power when the basic blocks reuse equal conflicts. The current preliminary prototype tool iterates over the basic blocks, using flow-graph balancing, to find an ordering. In a second phase it combines the partial conflict graph results of the different basic blocks.

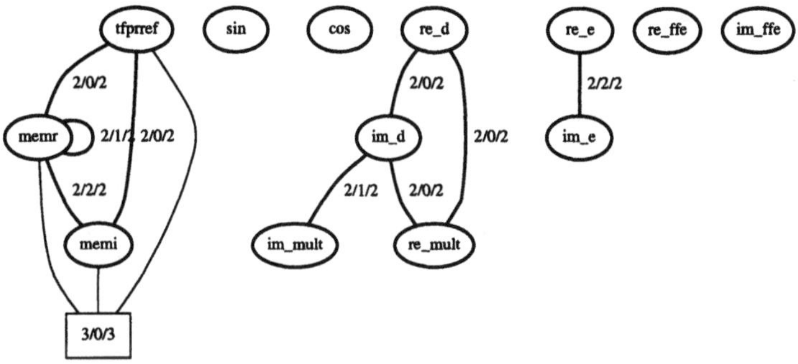

Figure 3.15 The conflict graph output of the storage cycle budget distribution step constrains the memory architecture design.

As output result of the storage cycle budget distribution step, an ECG is produced for a given cycle budget. Figure 3.15 shows an example ECG for the DAB synchro core, constraining the final memory architecture. All the arrays of the DAB synchro core are put in separate bubbles. Array *memr* has a self conflict forcing a dual-port memory. Eight array conflicts are introduced. For instance the conflict between *memr* and *memi* forcing them in two different memories or in a dual port memory. The box in Figure 3.15 represents a hyperedge between *memi*,*memr* and *tfprref*. In other words, a memory cycle exists where these three arrays are accessed simultaneously. Consequently these three signals cannot be stored in a single dual-port memory. The next step in the next section, generates a custom cost-optimized memory architecture based on these constraints.

3.9 MEMORY ALLOCATION AND ASSIGNMENT

Based on the extended conflict graph obtained with the cycle budget distribution step, the detailed memory organization can be decided. This task is tackled in the memory allocation and assignment step, for which a systematic technique has been published in [60] and in [9, chapter 11].

The memory allocation and assignment is done in three substeps. First, the number and type of memories is chosen, during the memory allocation substep. Then, every basic group is assigned to one of the allocated memories in the signal-to-memory assignment substep. Finally, the dimensions of each memory are determined by the dimensions of the basic groups assigned to it and the related bus and port organisation are derived. After these substeps, the memory architecture is fully determined. The actual data organisation inside each memory is still partly free however. That should be fixed only during the in-place mapping step described further on.

The memory architecture cost heavily hinges on the signal-to-memory assignment. Therefore, it is essential to make a good decision here. Because this is a non-trivial task when many basic groups are involved, we have developed software to automate it. The signal-to-memory assignment tool uses the cost models from a memory library to find the optimal assignment. The resulting cost figures implicitly give the designer feedback about the memory architecture cost.

The memory allocation freedom for on-chip memories can be used also to trade off between power consumption and chip area. The power consumption of on-chip memories is dependent in a sublinear way on the memory size. Therefore, one way to reduce the on-chip power consumption is to split up memories in smaller units. On the other hand, every additional memory usually incurs an overhead in area for the peripheral units. In addition, the complexity of the system busses usually increases by adding extra memories. These penalties have to be traded off with the power gain of additional memories by evaluating the cost of the overall memory architecture.

An optimal memory assignment for the DAB demonstrator is given in Figure 3.16 obeying the conflict graph of Figure 3.15. Two single port RAMs, one dual port RAM and one ROM are allocated. The black rectangulars are the arrays assigned to these memories. The width and height of the squares represent respectively the bit-width and address range[4]. The signals sin, cos and $tpfrref$ are read only arrays and forced in the ROM. This is possible because no conflict occurs between these signals. The minimum memory allocation, one large dual-port memory, ends up in a power costly implementation. Therefore two additional single port memories are allocated. It seems unlogical to store all the small signals in the dual-port memory. However, it cannot be avoided that the bit-width of the dual-port memory is large because re_d, im_d and re_mult are in conflict and need to be stored in three different memories. By putting the small signals in the dual-port memory, wide single port memories are avoided. Indeed, allocating an additional single port memory will lead to a lower power consumption (see Figure 3.17. The wide single-port memory consumes much less power then a wide dual-port memory. Our automated signal-to-memory assignment tool allows to carefully explore this search space.

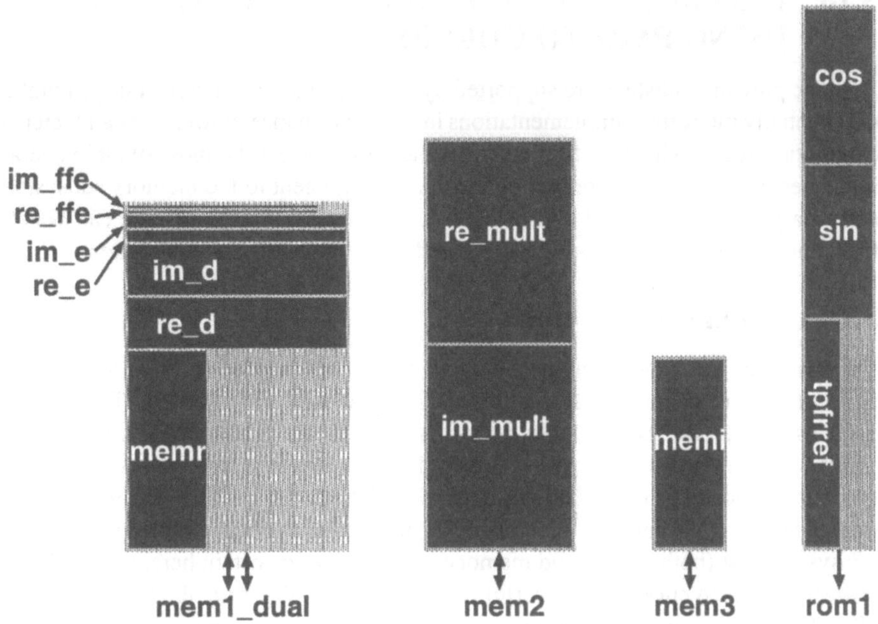

Figure 3.16 A memory architecture which obeys the conflict graph of Figure 3.15

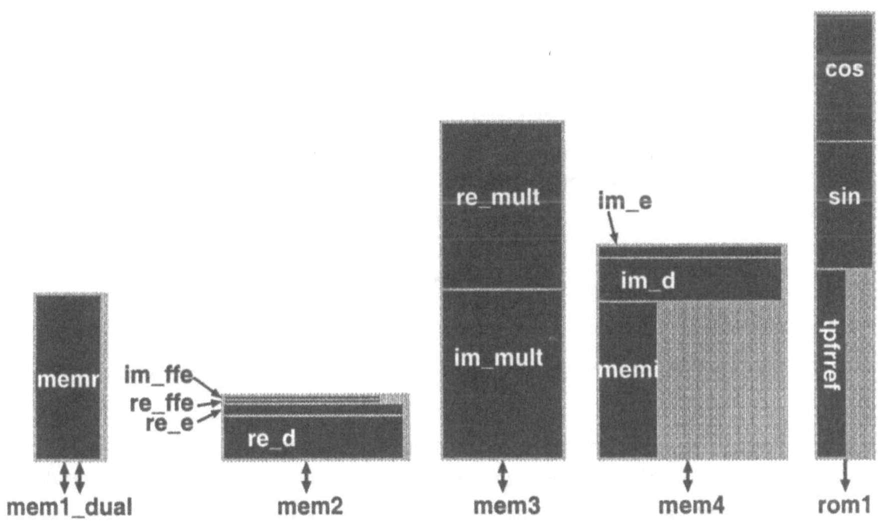

Figure 3.17 Allocating one additional single port memory reduces the power consumption.

3.10 GLOBAL PERFORMANCE-COST TRADE-OFF USING PARETO CURVES

Both the previous substeps are supported by tools. The tools generate a large number of potentially interesting implementations in the performance, power, area and latency exploration space. The use of Pareto curves allows to select the most suitable overall implementation. Tradeoffs regarding the cycle assignment to the memory subsystem and data-path are supported. Also the task to processor mapping stage can benefit from this information. This will be discussed in the remainder of this section.

3.10.1 General principles

The cost increases when a higher memory bandwidth is needed. The lower cycle budget the more memory accesses in parallel (=bandwidth) is needed. This section explains how to use the available trade-offs between cycle budget for a given task, and the memory cost related to it. This leads to the use of Pareto curves. The DAB synchro demonstrator is worked out in detail to illustrate this. The data transfer and storage related cycle budget ("memory cycle budget") is strongly coupled to the memory system cost (both energy and memory size/area are important here, as mentioned earlier). A Pareto curve is a powerful instrument to be able to make the right trade-offs. The Pareto curve only represents the potentially interesting points in a search space with multiple axes and excludes all the solutions which have an equally good or worse solution for all the axes.

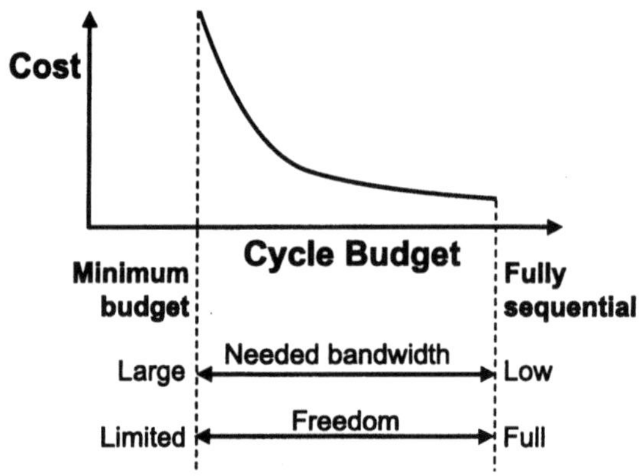

Figure 3.18 Pareto curve for trading of memory cycle budget vs. cost

In most cases, designers of real-time systems will assume that the time for executing a given (concurrently executed) task is predefined by "some" initial system decision step. Usually, this is only seemingly so though. When several communicating tasks or threads are co-operating concurrently, then trade-offs can be made between the cycle budget and timing constraints assigned to each of them, which makes the overall problem much more complex. Moreover, also in data-dominated applications the memory subsystem is not the only cost for the complete system. In particular after our DTSE optimisation steps, the data-path is typically not negligible any more afterwards, especially for instruction-set processor realisations.

The memory cost increases when lowering the cycle budget (see Figure 3.18). When lowering the cycle budget, a more energy consuming and more complex (larger) memory architecture is needed to deliver the required bandwidth. Note that the lower cycle budget is not obtained by reducing the number of memory accesses, as in the case of algorithmic changes and dataflow transformations. These higher-level code transformation related optimizations can be performed platform and cycle budget independent, prior to the trade-off which is the focus here. Therefore, during the currently discussed step in the system design trajectory, the amount of data transferred to the data-path remains equal over the complete cycle budget range. Nevertheless, some control flow transformations and data reuse optimizations can be beneficial for energy consumption and not for the cycle budget (or vice versa). These type of optimizations still have to be considered in the trade-off.

The interesting range of such a Pareto graph on the cycle budget axis, should be defined from the critical path up to the fully sequential memory access path. In the fully sequential case all the memory accesses can be transferred over a single port (lowest bandwidth). However, the number of memories is fully free and the memory architecture can be freely defined. Therefore, the energy is the lowest. In the other extreme, the critical path, only a limited memory access ordering is valid. Many memory accesses are performed in parallel. Consequently, the constraints on the memory system are large and a limited freedom is available. As a result, the energy consumption goes up. Due to the more restricted cycle budget, the power itself increases even more of course, but the latter involves a more traditional trade-off which is not our focus here. This energy-performance trade-off is now feasible on real-life applications thanks to the use of powerful novel tools which we have recently developed [1].

Figure 3.19 shows the trade-off for the complete cycle budget range of the DAB synchronisation core. This is obtained by letting the tools explore the cycle budget starting from the fully sequential mode, and then progress through the most interesting memory organisations (from a cost point of view) to reduce the cycle budget. The number of allocated memories is four for the entire graph shown in this figure, but it will be shown further on that an additional exploration can be added over that parameter, too. Originally, only single-port memories are employed. When a dual-ported memory has to be added, a clear discontinuity is present in the energy function. In order to reduce the budget below 85 kCycles, dual-ported memories are indeed needed. The allocation of 4 dual-ported memories allows to decrease the cycle budget close to the critical path (55047 cycles). In principle, the tool could also explore even more costly memory organisations with multi-port memories but the cost becomes so huge then (if they would even exist in a realistic memory library) and the gains in cycle budget become so small that it does not make any practical sense to continue at that stage.

In case of a single thread and the memory subsystem being energy dominant, the trade-off can be based solely on the memory organisation Pareto curve. The given cycle budget defines the best memory architecture and its corresponding energy and area cost. Due to the use of latency hiding and a "system-level pipeline" between

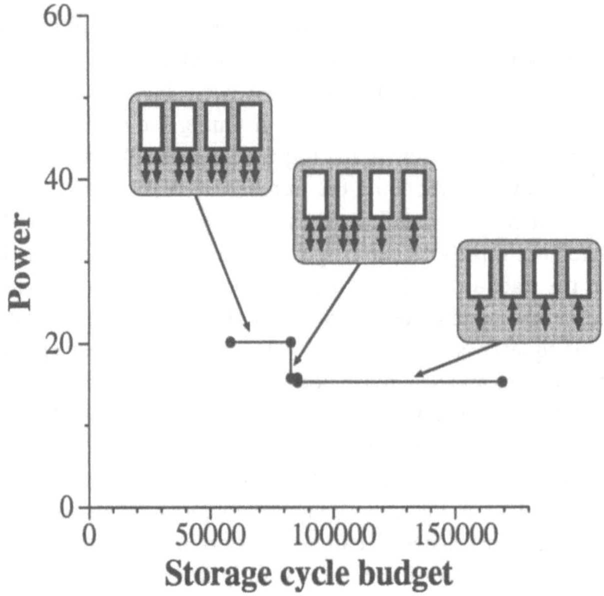

Figure 3.19 Pareto curve for Digital Audio Broadcast application.

the computation and data communication, most of the cycles where memory access takes place are also useful for the data-path. Still, to reduce the cost of the data-path realisation, it is important to leave some of the total cycle budget purely available for computation cycles (see [6]).

3.10.2 Application to task concurrency management exploration

The performance-energy function can be used even better to trade off cycles assigned to different concurrently operating tasks in a complex system, which are usually mapped to a multi-processor architecture. Our Pareto curves should indeed be generated per concurrently executed task. It is then clearly visible that assigning too few cycles to a single task causes the cost of the entire system to go up. The Pareto curves also show the minimal (realistic) cycle budget per task. The cycle and energy estimates can clearly help the designer to assign concurrent tasks to processors and to distribute the cycles within the processors over the different tasks (see Figure 3.20). Minimizing the overall energy within a heterogeneous multi-processor is then possible by applying Pareto function minimization on all the energy-cycle functions together. Even when multiple tasks have to be assigned to a single processor under control of an operating system, interesting trade-offs are feasible.

Next, we will give a detailed illustration of this in the context of a DAB receiver. The DAB synchronization core is divided in three co-operating tasks on which we will focus. First there is an FFT task, to be executed for every symbol of a frame (in Mode I, a frame contains 76 symbols). Concurrent with this FFT, the synchronization runs. It works with the first symbol of every frame, and consists of a correlation task

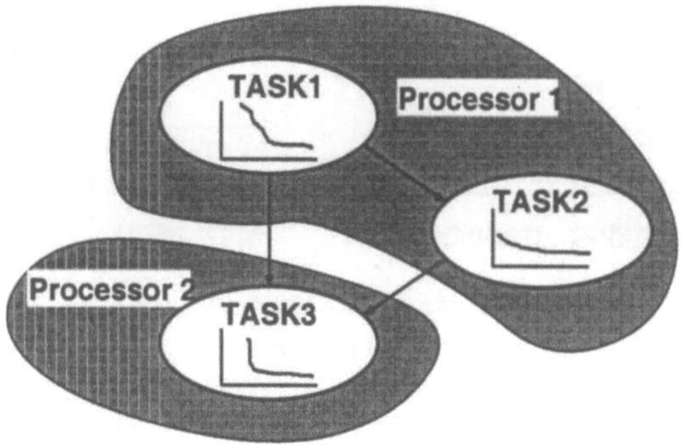

Figure 3.20 Tradeoff cycles assigned tasks.

followed by an Impulse Response Analysis (IR Analysis) task. An overall real-time constraint of 96ms, the frame length, applies for all three tasks together. Assuming a speed of 40MHz, a low-speed and low-energy mode, 3840K cycles are available in total. In addition, constraints exist due to the delay before the data comes available from the receiver signal processing.

A naive design would allocate a separate processor for the FFT tasks and for the two correlation tasks. This would also require a separate memory organization and communication channels between the two processors. The top figure in Figure 3.21 shows how the tasks are organized in time, and what the memory architecture is for this processor.

With the correlation done in parallel to the FFT, the processing and memory resources are very much underutilized. A designer can then decide to map all three tasks onto one processor, and to make use of the spare time in the NULL symbol to perform the correlation. However, it turns out that this leads to a very costly solution. A very high bandwidth to the memories is required to complete the correlation in time, and a six-port memory must be allocated to provide this bandwidth. The complexity of such a memory architecture makes this option simply infeasible.

Consider now the Pareto curves for the three DAB tasks, shown in Figure 3.22. The numbers 1 and 2 indicate the cycle budgets corresponding to the two alternatives above. Indeed, it can be seen from the Pareto curves that the memory organization cost is minimal for the relaxed cycle budgets of the first, two-processor solution. If the one-processor solution is chosen, where the two synchronization tasks are compressed into the smaller cycle budget of the NULL symbol, it turns out that the cycle budget is then in the very high cost range.

However, from the Pareto curves it is immediately obvious that there is a more promising alternative: choosing the points just before the cost breakpoints of the two correlator tasks, as indicated by the number 3 in Figure 3.22. The cycles required to avoid the second processor can be taken away from the FFT task, where also still some slack is available before the cost jump. An ordering of the tasks corresponding to this decision, and the associated memory architecture, are shown at the bottom part of Figure 3.21. Thus, by exploring the trade-off with the aid of the Pareto curves,

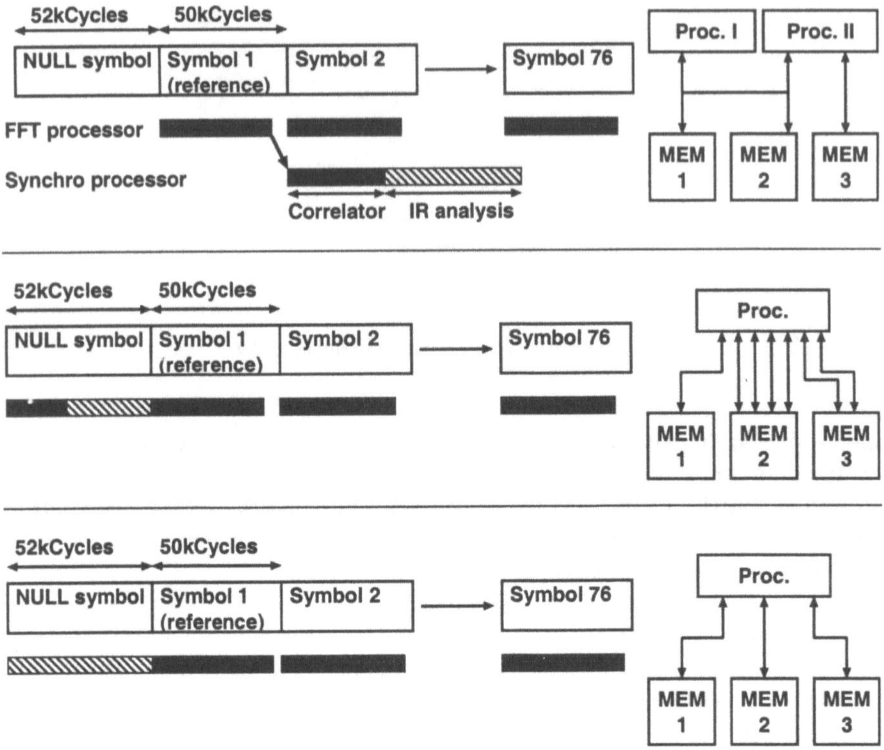

Figure 3.21 The original task to processor mapping needs 2 processors (top); A naive more energy consuming combination of the two tasks to a single processor (middle); An improved task rescheduling overcomes the cost increase (bottom).

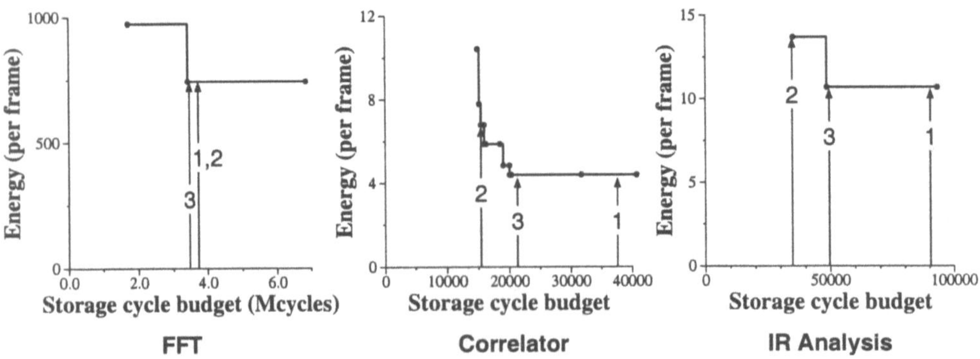

Figure 3.22 Pareto curves of three different subtasks of the DAB.

the global system cost is effectively reduced (from two processors to one), while still meeting the real-time constraint and without an energy penalty[5].

3.11 IN-PLACE MAPPING

This task is illustrated in Fig. 3.23. Initial methods have been investigated for deciding on in-place storage of multi-dimensional signals [63, 64, 62]. Recently, further experiments have been performed to allow even more reduced storage requirements and to identify extensions for predefined memory organisation as in programmable DSP's [17, 18, 19, 12]. This has lead to a solid theoretical foundation for the in-place mapping task and the development of promising heuristics to solve this very complex problem in a reasonable CPU time in a 2-stage approach for M-D signals [20, 24, 25]: intra-signal windowing [22], followed by inter-signal placement [23]. Inter-signal in-place allows to store two signals in a set of memory locations where, before this optimization, only one of the signals was located. This is only possible when the life-times of the signal do not fully overlap each other. When the same story is true for elements in one signal, we call it intra-signal in-place. This is only possible when the information of some elements is no longer needed or valid before others are created.

These techniques have been implemented in a prototype tool for evaluating in-place storage of multi-dimensional signals. Very promising results have been obtained with this approach for realistic applications [22, 23, 24, 25].

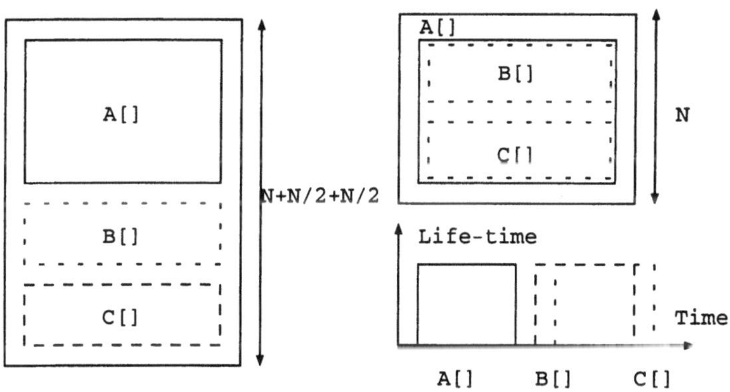

Figure 3.23 Illustration of in-place mapping task. In the worst-case, all signals require separate storage locations (left-hand side solution). However, if the life-times of signals $B[]$ and $C[]$ are not overlapping with the life-time of signal $A[]$, the space reserved in the memory for these groups can be shared (right-hand side, top).

Before giving an in-place example within the FFT context we present a new graphical representation of the FFT. The new representation includes the (time) ordering of the butterflies. The traditional representation (Figure 3.24 left) does not enforce butterfly execution ordering. In practice however, the traditional stage-sequential execution order is usually used. Our new representation introduces a time axis (Figure 3.24 right). Each line represents a butterfly and the end points of each line represent which data elements are used. So, the first line is the butterfly using element 0 and 8, followed by 1 and 9. This latter figure is completed with the traditional butterfly execution order.

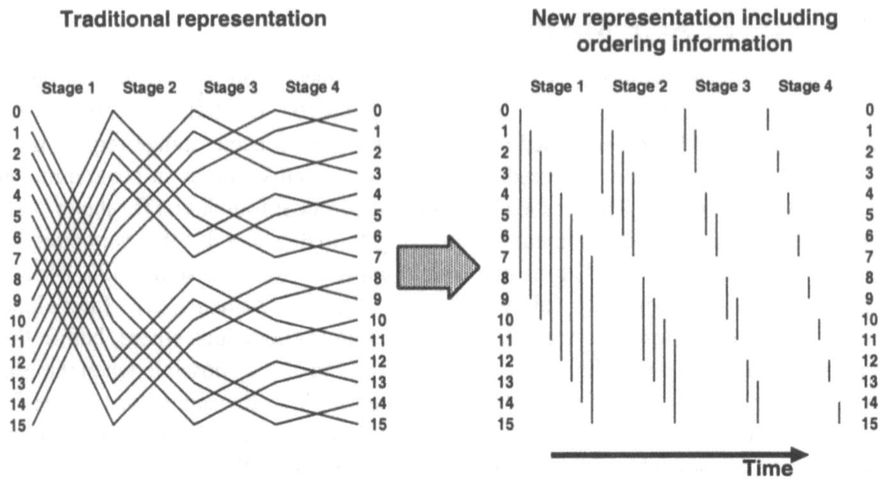

Figure 3.24 The new graphical representation for FFT includes the time ordering of the butterflies.

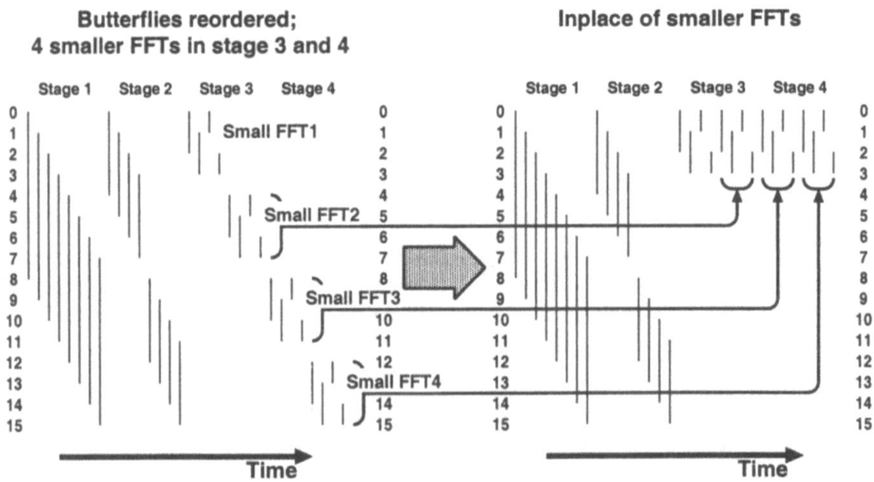

Figure 3.25 The butterflies of stage 3 and 4 are reordered towards 4 smaller FFTs. These smaller FFTs can be in-placed (reusing the same memory locations, here location 0—3).

In-place mapping is very useful in the context of an FFT, which is as such well-known. We will show however that the search space is larger than what is typically explored and that systematic exploration results can be obtained. A (large) FTT can be broken up in several smaller FFTs (as shown in Figure 3.25). The first two butterflies of stage 4 are moved close to the first 2 bufferflies of stage 3. Also, the other butterflies are reordered equally, mixing stage 3 and 4. The maximally used address space is reduced when executing these smaller FFTs sequentially (here maximally four elements are used). Due to the reduced address space, these FFTs can be in-placed in a smaller memory, consuming less power (here a four element memory).

The basic in-place principle can be applied to the FFT in many different ways. Indeed, this breakup can be made for every stage. Moreover, this can be done at multiple places causing a large exploration space. The power-gain does not come for free: the additional memory (and interconnect) causes an additional cost in terms of area. Fortunately, the useful exploration space for this optimization can be explored exhaustively using fast estimators. The power gain and memory overhead cost can be determined by feeding memory access counts to accurate memory models. Finally, the data can be collected in a graph as shown in Figure 3.26. This process can be largely automated with scripts. An easy trade-off can be made because the graph only includes the Pareto points (the Pareto concept is already explained in section 3.10).

Every point on the Pareto curve is another implementation. Four memory architecture implementation are pointed out to clearify the concept. Obviously, the implementation without memory hierarchy has the smallest area overhead and consumes most power[6] (leftmost point). This point is used as a reference for power and area. Since this memory space is needed anyway, the addition memory area (relative to this obligatory memory) is counted. The full hierarchy has the largest area and lowest power when ignoring the bus overhead (rightmost points). Two more interesting points in the middle are highlighted. The idea introduced in this section is not fundamentally new for instances of the FFT but apparently it has never been systematically and exhaustively explored before for a parameterized FFT core for the entire search space that is available.

Additionally, the FFT example show the big impact of the target platform and its cost models. Figure 3.26 shows the trade-off for using two different existing memory models. Due to the significant curve shape difference, another implementation in the Pareto space will be chosen as the optimal one for the different models.

3.12 HIGH-LEVEL ADDRESS OPTIMISATION AND ADDRESS UNIT SYNTHESIS

The stage to derive optimized address generation code from the initial high-level loop organisation, condition scopes and the array indices, is very crucial for data-dominated applications. The required steps are error-prone and tedious if performed manually for RMSP applications. Therefore, a methodology has been developed which is partly supported in IMEC's ADOPT environment [45, 47], which is embedded in a separate prototype tool-box. The subtasks addressed in this toolbox are complementary to the traditional high-level synthesis techniques as employed e.g. in Synopsys BC, so ADOPT can be seen as a high-level preprocessing stage.

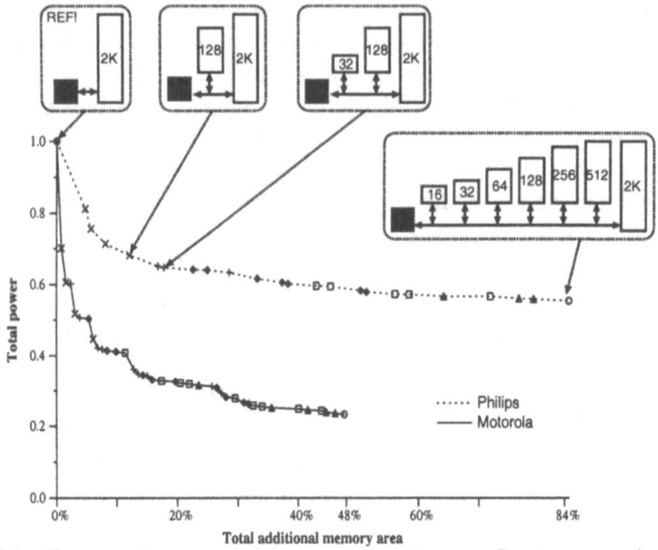

Figure 3.26 The power improvement versus the area increase Pareto curve when exploring all possibilities for in-placing a 2K FFT.

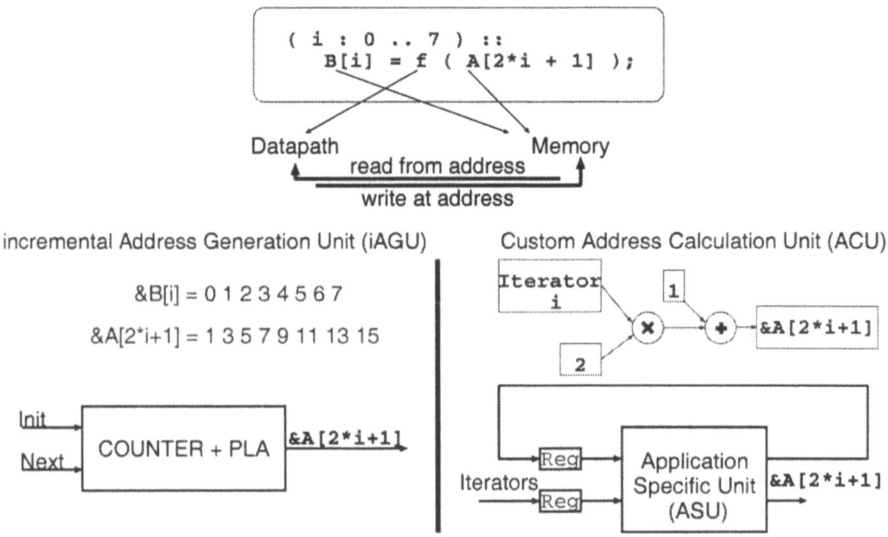

Figure 3.27 Illustration of address optimisation and hardware mapping task. After algebraic optimisation of the address expressions, they are efficiently shared on hardware units within the available cycle budget. Two target architecture styles are supported: logic/counter based or custom data-path (ASU) based. In the former case, address sequences are generated and realized as a counter modified by a two- or multi-level logic filter. In the ASU case, the address expressions are realized with custom arithmetic building blocks selected from a library.

The methodology is illustrated in Fig. 3.27 [47]. A prototype tool GATE has been developed for optimally transforming the (address) arithmetic operations in flow-graph expressions [32]. A model and automated steering techniques have been developed which allow to minimize the weighted operation cost, allowing invariance to structural changes in the specification. Other prototype tools allow merging [32, 33] and/or sharing [28] M-D index expressions such that the address cost is reduced [44]. This includes a measure for selecting the target architecture, i.e. logic/counter based [61, 46] or custom data-path (ASU) based [28, 29, 30, 31, 49, 50].

The effective usage of the Adopt method will be shown on the correlator of the DAB synchro core. The correlator discussed in section 3.6 is more complex, as shown in Figure 3.28. Three different functions ($funcA$, $funcB$ and $funcC$) are called repeatedly in the inner loop. Therefore, the loop interchange proposed in section 3.6 is not trivial. The remainder of this section shows how to obtain the goal in three steps. Firstly, introduce the correct amount of overhead in the pruning step for easy transformation. Secondly, apply the transformation. Thirdly, remove the overhead and improve the performance even further. These three steps are considered in the next paragraphs.

The original correlator (see Figure 3.28 is pruned and preprocessed to simplify the transformations. The loop body of the original description can be seen, from DTSE point of view, as an unrolled loop of three iterations. The three successive reads are the only important code for DTSE. A "Z" loop can be created when rolling the loop again. Additional conditions handle correct execution of the data-path issues. Indeed, the entire part is situated at level-3 and can be pruned out for DTSE. The last preprocessing step merges the y and z to a single yz loop. Note the added computational expensive module (%) operations.

Due to the preprocessing, the loop interchange transformation becomes nearly trivial (see also section 3.6). As a result, the range of the yz loop is extended to $M + N$ and the inner x loop bounds are dependent on the yz loop (S(yz)..E(yz)). We don not want to go into details how these loop bounds can be obtained formally [36].

The DTSE stage has optimized the data memory organisation but it has a very negative impact on performance on a programmable processor, because a complex control and addressing scheme has been introduced. Our Adopt methodology allows to remove this overhead again. Moreover, the combined methodologies will typically lead to an improved performance over the original specification because the data bottleneck is removed. The first substep of the Adopt script minimizes the algebraic calculation cost. This includes global scope common subexpression elimination, constant propagation and global loop invariant code hoisting. A simple example can be found in the 4th cote Figure 3.28 where the three times repeated subexpression "$((yz - x)\%3)$" can be executed only once.

The next Adopt optimization substep is non-linear strength reduction, where costly instructions are replaced by less costly instructions. A modulo is a cost-inefficient instruction on a typical processor. Typically however, the modulo operand tmp_yz will change only with one with respect to the previous calculation. Therefore an incremental solution can be used in our context, for which the modulo operation is replaced by a condition and decrement (see 5th code of Figure 3.28). Only a single modulo operation is needed for the initialisation before the loop start. Indeed, this modulo can be moved higher up in the loop nest using the same approach again. Finally, all iterator modulo operators can be removed. The so far mentioned code optimisations are processor-independent and have no penalty in performance nor power. The next optimization is processor-dependent and cannot be performed to just any processor without increasing the global cost. Accurate cost calculation/estimation is needed to trade off the costs involved.

The remaining bottleneck is due to the conditional statements in the loop body. Remember that they appeared when rolling up the "z" loop. Basically, this loop should be unrolled again in the "x" direction as shown in the last code of Figure 3.28. Unfortunately, the code is much less regular, enforcing a special behaviour at the edges. We skip the preamble and postamble, focusing on the main part (most executed) of the algorithm. The calculation is column per column from top to bottom. The execution order is therefore $funcC$, $funcB$, $funcC$ (note this is in reverse order compared to the original description). However, every column does not necessarily start with a $funcC$. Therefore, a column-dependent preamble and postamble are needed (shaded areas). The most executed [7] *unrolled* inner loop is denoted with the black box. On one hand the conditions are not needed in the inner loop, which speeds up this implementation. On the other hand, the increased code size may have a negative impact on power (larger code ROM). Moreover, instruction cache misses might cause a countereffect on performance. So this choice belongs to the processor-specific steps.

3.13 CONCLUSIONS

In this chapter, we have illustrated IMEC's systematic processor-level DTSE methodology [9, 10]. It starts from a system specification with accesses on multi-dimensional (M-D) signals which can be statically ordered. The output is an optimized memory organisation combined with a transformed specification which is the input for the architecture (high-level) synthesis when custom realizations are envisioned, or for the software compilation stage in the case of predefined processors. The most time-consuming and error-prone steps within this methodology are becoming supported by tools, developed in the context of the ATOMIUM system exploration environment. We have also demonstrated the impact and usefulness of this methodology and the supporting tools on part of the recent Digital Audio Broadcast standard. The data transfer and storage related energy consumption is significantly reduced as a result, with about a factor 4 compared to the initial reference design.

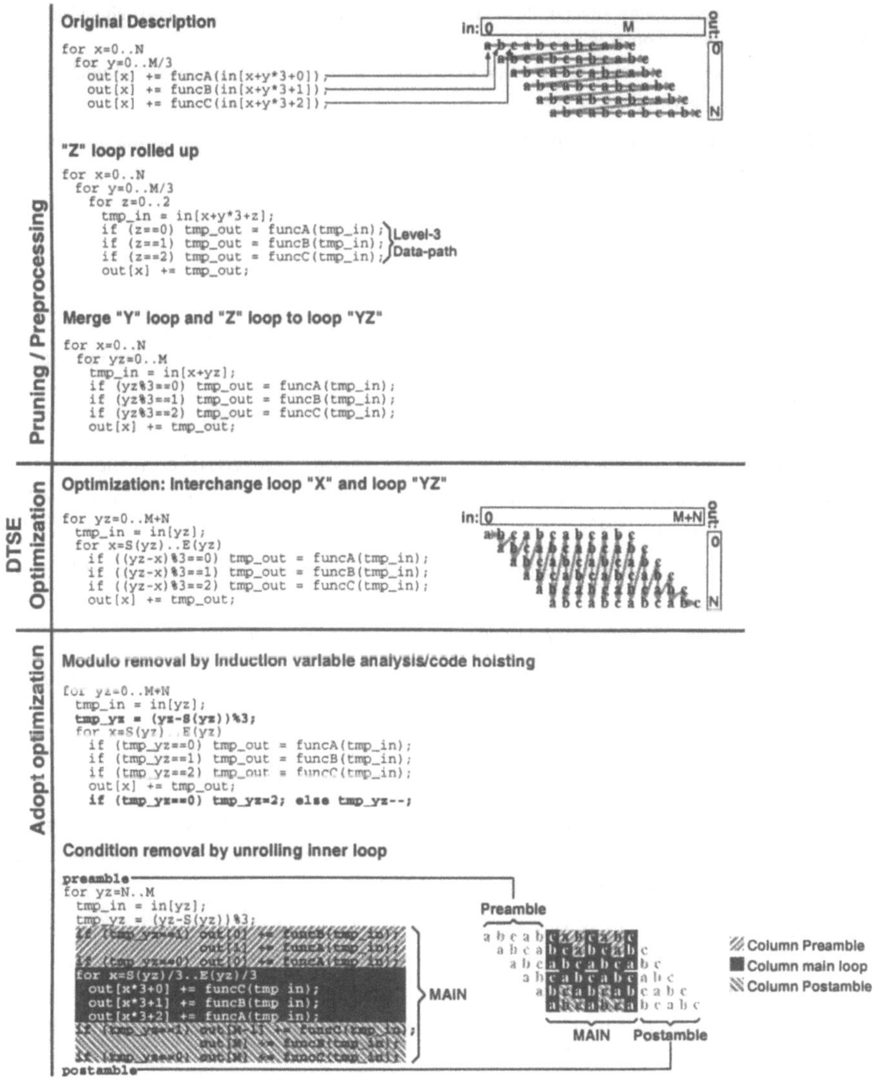

Figure 3.28 Address/control optimization steps to gain performance (ADOPT).

Notes

1. All of these prototype tools operate on models which allow run-time complexities which are dependent in a limited way on system parameters like the size of the loop iterators, as opposed to the scalar-based methods published in conventional high-level synthesis literature.

2. This order is extracted from the dependencies which reflect the outcome of each stage on the other stages. This enabling/disabling effect of the transformations is crucial to investigate in-depth before deciding on the most promising order in a script

3. The real implementation is more complex but cannot be explained because of confidential reasons.

4. The sizes are logarithmic for overview reasons.

5. It is assumed that no energy saving is possible by parallelizing the data-path part of the processors here, because they are already operating at the lowest Vdd with a low clock frequency.

6. Assuming single port memories only.

7. The dominance of the inner loop is not visible due to scaling of the example.

References

[1] More information and related papers on the Acropolis project are available at the IMEC web site:
`http://www.imec.be/acropolis/Welcome.html.`

[2] S.Amarasinghe, J.Anderson, M.Lam, and C.Tseng, "The SUIF compiler for scalable parallel machines", in *Proc. of the 7th SIAM Conf. on Parallel Proc. for Scientific Computing*, 1995.

[3] More information and related papers on the Atomium project are available at the IMEC web site:
`http://www.imec.be/atomium/Welcome.html.`

[4] B.M. Baas, "An Energy-Efficient Single-Chip FFT Processor" in *IEEE Symp. on VLSI*, Honolulu HI, 13-15 june 1996.

[5] U.Banerjee, R.Eigenmann, A.Nicolau, D.Padua, "Automatic program parallelisation", *Proc. of the IEEE*, invited paper, Vol.81, No.2, Feb. 1993.

[6] E.Brockmeyer, S.Wuytack, A.Vandecappelle, F.Catthoor, "Low power storage for hierarchical graphs", *Proc. 3rd ACM/IEEE Design and Test in Europe Conf.*, Paris, France, User-Forum pp.249-254, April 2000.

[7] F.Catthoor, F.Franssen, S.Wuytack, L.Nachtergaele, H.De Man, "Global communication and memory optimizing transformations for low power signal processing systems", *IEEE workshop on VLSI signal processing*, La Jolla CA, Oct. 1994. Also in *VLSI Signal Processing VII*, J.Rabaey, P.Chau, J.Eldon (eds.), IEEE Press, New York, pp.178-187, 1994.

[8] F.Catthoor, M.Janssen, L.Nachtergaele, H.De Man, "System-level data-flow transformation exploration and power-area trade-offs demonstrated on video coders", special issue on "Systematic trade-off analysis in signal processing systems design" (eds. M.Ibrahim, W.Wolf) in *Journal of VLSI Signal Processing*, Vol.18, No.1, Kluwer, Boston, pp.39-50, 1998.

[9] F.Catthoor, S.Wuytack, E.De Greef, F.Franssen, L.Nachtergaele. H.De Man, "System-level transformations for low power data transfer and storage", in paper

collection on "Low power CMOS design" (eds. A.Chandrakasan, R.Brodersen), IEEE Press, pp.609-618, 1998.

[10] F.Catthoor, S.Wuytack, E.De Greef, F.Balasa, P.Slock, "System exploration for custom low power data storage and transfer", chapter in "Digital Signal Processing for Multimedia Systems" (eds. K.Parhi, T.Nishitani), Marcel Dekker, Inc., New York, 1998.

[11] F.Catthoor, S.Wuytack, E.De Greef, F.Balasa, L.Nachtergaele, A.Vandecappelle, "Custom Memory Management Methodology – Exploration of Memory Organisation for Embedded Multimedia System Design", ISBN 0-7923-8288-9, Kluwer Acad. Publ., Boston, 1998.

[12] M.Cupak, F.Catthoor, "Efficient functional validation of system-level loop transformations for multi-media applications", *Proc. Electronic Circuits and Systems Conference*, Bratislava, Slovakia, pp.39-43, Sep. 1997.

[13] M.Cupak, C.Kulkarni, F.Catthoor, H.De Man, "Functional Validation of System-level Loop Transformations for Power Efficient Caching" *Proc. Wsh. on System Design Automation*, Dresden, Germany, March, 1998.

[14] F.Catthoor, "Energy-delay efficient data storage and transfer architectures: circuit techno logy versus design methodology solutions", *Proceedings of DATE'98*, Feb.23-25 1998.

[15] K.Danckaert, F.Catthoor, H.De Man, "Platform independent data transfer and storage exploration illustrated on a parallel cavity detection algorithm", *Proc. ACM Conf. on Par. and Dist. Proc. Techniques and Applications*, PDPTA'99, Vol.III, pp.1669-1675, Las Vegas NV, June 1999.

[16] K.Danckaert, F.Catthoor, H.De Man, "A loop transformation approach for combined parallelization and data transfer and storage optimization", accepted for *Proc. ACM Conf. on Par. and Dist. Proc. Techniques and Applications*, PDPTA'00, Las Vegas NV, June 2000.

[17] E.De Greef, F.Catthoor, H.De Man, "A memory-efficient, programmable multiprocessor architecture for real-time motion estimation type algorithms", *Intnl. Workshop on Algorithms and Parallel VLSI Architectures*, Leuven, Belgium, August 1994. Also in "Algorithms and Parallel VLSI Architectures III" (eds. M.Moonen, F.Catthoor), Elsevier, pp.191-202, 1995.

[18] E.De Greef, F.Catthoor, H.De Man, "Mapping real-time motion estimation type algorithms to memory-efficient, programmable multi-processor architectures", *Microprocessors and Microprogramming*, special issue on "Parallel Programmable Architectures and Compilation for Multi-dimensional Processing" (eds. F.Catthoor, M.Moonen), Elsevier, pp.409-423, Oct. 1995.

[19] E.De Greef, F.Catthoor, H.De Man, "Memory organization for video algorithms on programmable signal processors", *Proc. IEEE Int. Conf. on Computer Design*, Austin TX, pp.552-557, Oct. 1995.

[20] E.De Greef, F.Catthoor, H.De Man, "In-Place mapping and its relation to loop parallelisation", presented at *Dagstuhl on Loop Parallelisation*, Schloss Dagstuhl, Germany, April 1996.

[21] E.De Greef, "Combined flow-graph and polyhedral model", Internal Research Presentation, IMEC, May 1996.

[22] E.De Greef, F.Catthoor, H.De Man, "Memory Size Reduction through Storage Order Optimization for Embedded Parallel Multimedia Applications", *Intnl. Parallel Proc. Symp.(IPPS)* in Proc. Workshop on "Parallel Processing and Multimedia", Geneva, Switzerland, pp.84-98, April 1997.

[23] E.De Greef, F.Catthoor, H.De Man, "Array Placement for Storage Size Reduction in Embedded Multimedia Systems", *Proc. Intnl. Conf. on Applic.-Spec. Array Processors*, Zurich, Switzerland, pp.66-75, July 1997.

[24] E.De Greef, F.Catthoor, H.De Man, "Memory Size Reduction through Storage Order Optimization for Embedded Parallel Multimedia Applications", special issue on "Parallel Processing and Multi-media" (ed. A.Krikelis), in *Parallel Computing* Elsevier, Vol.23, No.12, Dec. 1997.

[25] E.De Greef, "Storage size reduction for multimedia applications", *Doctoral dissertation*, ESAT/EE Dept., K.U.Leuven, Belgium, Jan. 1998.

[26] J.Z.Fang, M.Lu, "An iteration partition approach for cache or local memory thrashing on parallel processing", *IEEE Trans. on Computers*, Vol.C-42, No.5, pp.529-546, May 1993.

[27] P.Feautrier, "Compiling for massively parallel architectures: a perspective", *Intnl. Workshop on Algorithms and Parallel VLSI Architectures*, Leuven, Belgium, August 1994. Also in "Algorithms and Parallel VLSI Architectures III" (eds. M.Moonen, F.Catthoor), Elsevier, pp.259-270, 1995.

[28] W.Geurts, F.Catthoor, H.De Man, "Heuristic techniques for the synthesis of complex functional units", *Proc. 4th ACM/IEEE Europ. Design Automation Conf.*, Paris, France, pp.552-556, Feb. 1993.

[29] W.Geurts, F.Catthoor, H.De Man, "Quadratic Zero-one Programming-based Synthesis of Application-Specific Data Paths", *IEEE Trans. on Comp.-aided Design*, Vol.CAD-14, No.1, pp.1-11, Jan. 1995.

[30] W.Geurts, "Synthesis of accelerator data-paths for high-throughput signal processing applications", *Doctoral dissertation*, ESAT/EE Dept., K.U.Leuven, Belgium, March 1995.

[31] W.Geurts, F.Catthoor, S.Vernalde, H.De Man, "Accelerator data-paths synthesis for high-throughput signal processing applications", Kluwer Academic Publishers, Boston, 1996.

[32] J.M.Janssen, F.Catthoor, H.De Man, "A Specification Invariant Technique for Operation Cost Minimisation in Flow-graphs". *Proc. 7th ACM/IEEE Int. Workshop on High-Level Synthesis*, Niagara-on-the-Lake, Canada, pp.146-151, May 1994.

[33] J.M.Janssen, F.Catthoor, H.De Man, "A Specification Invariant Technique for Regularity Improvement between Flow-Graph Clusters", *Proc. European Design Automation Conf.*, Paris, France, pp.138-143, Feb. 1996.

[34] M.Jimenez, J.Llaberia, A.Fernandez, E.Morancho, "A unified transformation technique for multi-level blocking" *Proc. EuroPar Conference*, Lyon, France, August 1996. "Lecture notes in computer science" series, Springer Verlag, pp.402-405.

[35] D.Kolson, A.Nicolau, N.Dutt, "Minimization of memory traffic in high-level synthesis", *Proc. 31st ACM/IEEE Design Automation Conf.*, San Diego, CA, pp.149-154, June 1994.

[36] D.Kulkarni, M.Stumm, R.Unrau, "Implementing flexible computation rules with subexpression-level loop transformations", Technical report, Comp. Systems Res. Inst. Univ. of Toronto, Canada, 1995.

[37] van de Laar, F., Philips, N., and Olde Dubbelink, R., "General-purpose and application-specific design of a DAB channel decoder", *EBU Technical Review*, 258 (Winter 1993), 25 – 35, ISSN 1019-6587.

[38] Langen, E., "The Philips DAB 452 test receiver", *DAB Newsletter*, F. Kozamernik, Ed., no. 6. European Broadcasting Union, autumn 1994, pp. 6–10.

[39] W.Li, K.Pingali. "A singular loop transformation framework based on non-singular matrices", *Proc. 5th Annual Workshop on Languages and Compilers for Parallelism*, New Haven CN, August 1992.

[40] P.Lippens, J.van Meerbergen, W.Verhaegh, A.van der Werf, "Allocation of multiport memories for hierarchical data streams", *Proc. IEEE Int. Conf. Comp. Aided Design*, Santa Clara CA, Nov. 1993.

[41] N.Manjiakian, T.Abdelrahman, "Reduction of cache conflicts in loop nests", Technical report CSRI-318, Comp. Systems Res. Inst. Univ. of Toronto, Canada, March 1995.

[42] K.Masselos, F.Catthoor, C.E.Goutis, H.DeMan, "Low Power Mapping of Video Processing Applications on VLIW Multimedia Processors", *IEEE Alessandro Volta Memorial Intnl. Wsh. on Low Power Design (VOLTA)*, Como, Italy, pp.52-60, March 1999.

[43] K.Masselos, F.Catthoor, C.E.Goutis, H.De Man, "Code size effects of power optimizing code transformations for embedded multimedia applications", *Proc. IEEE Wsh. on Power and Timing Modeling, Optimization and Simulation (PATMOS)*, Kos, Greece, pp.61-70, Oct. 1999.

[44] M.Miranda, F.Catthoor, H.De Man, "Address equation optimization and hardware sharing for real-time signal processing applications", *IEEE workshop on VLSI signal processing*, La Jolla CA, Oct. 1994. Also in *VLSI Signal Processing VII*, J.Rabaey, P.Chau, J.Eldon (eds.), IEEE Press, New York, pp.208-217, 1994.

[45] M.Miranda, F.Catthoor, M.Janssen, H.De Man, "ADOPT: Efficient Hardware Address Generation in Distributed Memory Architectures", *Proc. 9th ACM/IEEE Intnl. Symp. on System-Level Synthesis*, La Jolla CA, pp.20-25, Nov. 1996.

[46] M.Miranda, M.Kaspar, F.Catthoor, H.De Man, "Architectural Exploration And Optimization for Counter Based Hardware Address Generation", *Proc. European Design Automation Conf.*, Paris, France, pp.293-298, Feb. 1997.

[47] M.Miranda, F.Catthoor, M.Janssen, H.De Man, "High-level Address Optimisation and Synthesis Techniques for Data-Transfer Intensive Applications", *IEEE Trans. on VLSI Systems*, Vol.6, No.4, pp.677-686, Dec. 1998.

[48] L.Nachtergaele, F.Catthoor, F.Balasa, F.Franssen, E.De Greef, H.Samsom, H.De Man, "Optimization of memory organization and hierarchy for decreased size and power in video and image processing systems", *Proc. Intnl. Workshop on Memory Technology, Design and Testing*, San Jose CA, pp.82-87, Aug. 1995.

[49] S.Note, W.Geurts, F.Catthoor, H.De Man, "Cathedral III: Architecture driven high-level synthesis for high throughput DSP applications", *Proc. 28th ACM/IEEE Design Automation Conf.*, San Francisco CA, pp.597-602, June 1991.

[50] S.Note, "Mapping high throughput signal processing algorithms into dedicated data-path architectures" *Doctoral dissertation*, ESAT/EE Dept., K.U.Leuven, Belgium, March 1991.

[51] D.A.Padua, M.J.Wolfe. "Advanced compiler optimizations for supercomputers", *Communications of the ACM*, Vol.29, No.12, pp.1184-1201, 1986.

[52] P.R.Panda, N.D.Dutt, A.Nicolau, "Memory issues in embedded in systems-on-chip: optimization and exploration", Kluwer Acad. Publ., Boston, 1999.

[53] L.Ramachandran, D.Gajski, V.Chaiyakul, "An algorithm for array variable clustering", *Proc. 5th ACM/IEEE Europ. Design and Test Conf.*, Paris, France, pp.262-266, Feb. 1994.

[54] H.Samsom, L. Claesen, H. De Man, "SynGuide: an environment for doing interactive correctness preserving transformations", *IEEE workshop on VLSI signal processing*, Veldhoven, The Netherlands, Oct. 1993. Also in *VLSI Signal Processing VI*, L.Eggermont, P.Dewilde, E.Deprettere, J.van Meerbergen (eds.), IEEE Press, New York, pp.269-277, 1993.

[55] H.Samsom, F.Franssen, F.Catthoor, H.De Man, "Verification of loop transformations for real time signal processing applications", *IEEE workshop on VLSI signal processing*, La Jolla CA, Oct. 1994. Also in *VLSI Signal Processing VII*, J.Rabaey, P.Chau, J.Eldon (eds.), IEEE Press, New York, pp.269-277, 1994.

[56] H.Samsom, F.Franssen, F.Catthoor, H.De Man, "System-level Verification of Video and Image Processing Specifications", *Proc. 8th ACM/IEEE Intnl. Symp. on System-Level Synthesis*, Cannes, pp.144-149, Sep. 1995.

[57] H.Samsom, "Formal verification and transformation of video and image processing applications", *Doctoral dissertation*, ESAT/EE Dept., K.U.Leuven, Belgium, Oct. 1995.

[58] H.Schmidt and D.Thomas. "Address generation for memories containing multiple arrays", *Proc. IEEE Int. Conf. Comp. Aided Design*, San Jose CA, pp.510–514, Nov. 1995.

[59] O.Sentieys, D.Chillet, J.P.Diguet, J.Philippe, "Memory module selection for high-level synthesis", *Proc. IEEE workshop on VLSI signal processing*, Monterey CA, Oct. 1996.

[60] P.Slock, S.Wuytack, F.Catthoor, G.de Jong, "Fast and extensive system-level memory exploration for ATM applications", *Proc. 10th ACM/IEEE Intnl. Symp. on System-Level Synthesis*, Antwerp, Belgium, pp.74-81, Sep. 1997.

[61] B.Vanhoof, M.Kaspar, P.Schaumont, "Address generation within Cathedral-2/3", SPRITE deliverable report C3.g/IMEC/Y5m12/1, Dec. 1993.

[62] J.Vanhoof, I.Bolsens, H.De Man, "Compiling multi-dimensional data streams into distributed DSP ASIC memory", *Proc. IEEE Int. Conf. Comp. Aided Design*, Santa Clara CA, pp.272-275, Nov. 1991.

[63] I.Verbauwhede, F.Catthoor, J.Vandewalle, H.De Man, "High-level memory management for real-time signal processing of algebraic algorithms on application-specific micro-coded processors", *Proc. Intnl. Workshop on Algorithms and Parallel VLSI Architectures*, Pont-a-Mousson, France, June 1990. Also in *Algorithms and Parallel VLSI Architectures*, Vol.B, E.Deprettere, A.Van der Veen (eds.), Elsevier, Amsterdam, pp.353-362, 1991.

[64] I.Verbauwhede, F.Catthoor, J.Vandewalle, H.De Man, "In-place memory management of algebraic algorithms on application-specific IC's", *Journal of VLSI signal processing*, Vol.3, Kluwer, Boston, pp.193-200, 1991.

[65] F.Vermeulen, F.Catthoor, D.Verkest, H.De Man, "A System-Level Reuse Methodology for Embedded Data-Dominated Applications", *Proc. IEEE Wsh. on Signal Processing Systems (SIPS)*, Boston MA, IEEE Press, pp.551-560, Oct. 1998.

[66] M.van Swaaij, F.Franssen, F.Catthoor, H.De Man, "Modelling data and control flow for high level memory management", *Proc. 3rd ACM/IEEE Europ. Design Automation Conf.*, Brussels, Belgium, pp.8-13, March 1992.

[67] M.van Swaaij, F.Franssen, F.Catthoor, H.De Man, "Automating high-level control flow transformations for DSP memory management", *Proc. IEEE workshop on VLSI signal processing*, Napa Valley CA, Oct. 1992. Also in *VLSI Signal Processing V*, K.Yao, R.Jain, W.Przytula (eds.), IEEE Press, New York, pp.397-406, 1992.

[68] S.Wuytack, F.Catthoor, L.Nachtergaele, H.De Man, "Power Exploration for Data Dominated Video Applications", *Proc. IEEE Intnl. Symp. on Low Power Design*, Monterey, pp.359-364, Aug. 1996.

[69] S.Wuytack, F.Catthoor, L.Nachtergaele, H.De Man, "Power Exploration for Data Dominated Video Applications", *Proc. IEEE Intnl. Symp. on Low Power Design*, Monterey, pp.359-364, Aug. 1996.

[70] S.Wuytack, F.Catthoor, G.De Jong, B.Lin, H.De Man, "Flow Graph Balancing for Minimizing the Required Memory Bandwidth", *Proc. 9th ACM/IEEE Intnl. Symp. on System-Level Synthesis*, La Jolla CA, pp.127-132, Nov. 1996.

[71] S.Wuytack, J.P.Diguet, F.Catthoor, H.De Man, "Formalized methodology for data reuse exploration for low-power hierarchical memory mappings", *IEEE Trans. on VLSI Systems*, Vol.6, No.4, pp.529-537, Dec. 1998.

4 HIGH-LEVEL POWER ESTIMATION METHODOLOGY APPLIED FOR PROCESSOR-LEVEL DTSE

Paul Lippens, Natalino Busa, Jos Huisken, Rafael Peset Llopis

Philips Research Laboratories Eindhoven

Paul.Lippens@philips.com

The huge integration capability of modern technologies allows very complex systems on a single IC. The design of such ICs is a very complex task due to the challenge of managing their complexity. There are two approaches to deal with complexity. Firstly, to increase the level of abstraction by using architectural synthesis tools. Secondly, to partition these ICs into several different types of sub-designs, such as embedded microprocessors and DSPs. Figure 4.1 gives an overview of a typical digital design flow. A behavioural VHDL description is made of the design, based on the specifications. This description is partitioned into several sub-designs, which are translated by several different architectural synthesis tools, and/or manually by a designer into a Register-Transfer level (RTL) VHDL description. The logic synthesis tool transforms the latter into a VHDL gate netlist, which is the input to layout synthesis.

A very important observation is that the higher the level of abstraction, the larger the impact of the design decisions on the performance of the design. For example, behavioural level decisions are likely to have a larger impact on the power dissipation than layout level decisions. Behavioral and RTL power estimation tools therefore are essential for the design of complex low-power ICs. However, most power estimation effort has been dedicated to the netlist and layout levels. Since power estimation at higher levels of abstraction is relatively a new research domain, a lot of work has to be done to develop feasible solutions.

In this chapter we address the issue of high-level power estimation. In Section 4.1, a new technique for power estimation and analysis based on high-level (RT-level and above) descriptions of digital circuits in presented. In Section 4.2, this approach is applied in the context of a digital chip for the demodulation and the decoding of DAB (Digital Audio Broadcast) signals with special emphasis on the processor-level DTSE context.

F. Catthoor (ed.),
Unified Low-power Design Flow for Data-dominated Multi-media and Telecom Applications, 65-96.

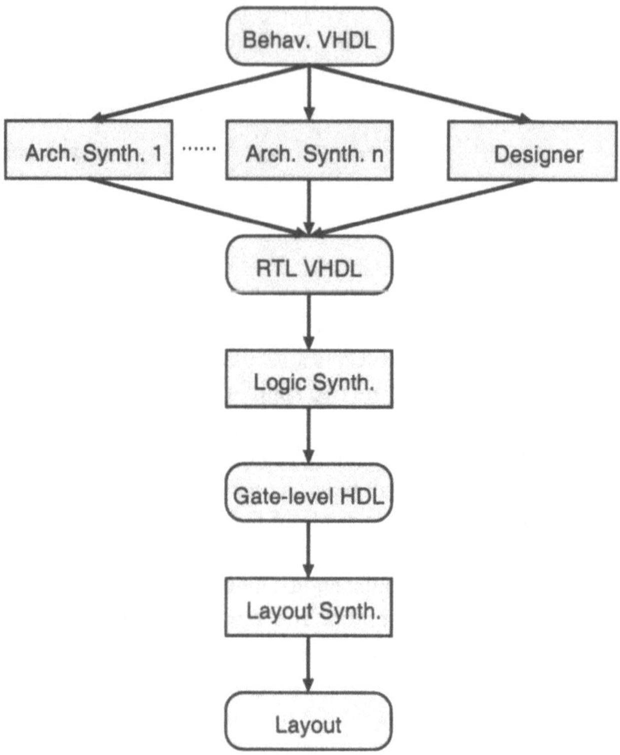

Figure 4.1 A typical digital design flow.

4.1 HIGH-LEVEL POWER ESTIMATION

In this section, a new approach in high-level power estimation is presented, which can be used both at the RT and behavioural levels. We will emphasize on RTL power estimation. However, the same approach can also be used at the behavioural level.

The outline of this section is as follows: Section 4.1.1 analyzes the shortcomings of the previously published power estimation approaches, and gives a list of requirements for RTL power estimation. Section 4.1.2 presents our RT and behavioural level power estimation approach. In Section 4.1.3 the proposed approach is benchmarked to gate-level power estimation.

4.1.1 Power Estimation

The dominant source of power dissipation in CMOS circuits is the charging and discharging of the capacitances [22]. This capacitive dissipation (P) is given by:

$$P = \sum_i \alpha_i \times C_i \times V_{dd}^2 \times f_{clk}$$

Here, α_i stands for the switching activity, C_i for the node capacitance, V_{dd} for the power supply voltage, and f_{clk} for the clock frequency. Furthermore, i encompasses all nodes of the circuit.

4.1.1.1 Power Estimation Requirements. In this section, a number of requirements with respect to high-level power estimation in digital circuits are stated.

- Power estimates have to be extracted from a description of the circuit behaviour (RTL or behavioural).
- It must be able to cope with 'any' VHDL description of digital circuitry, synthesizable or not.
- An important question is what accuracy is required for RTL power estimation. The absence of many implementation details in RTL descriptions makes the computation of accurate power estimates impossible. However, the good news is that the power optimization possibilities at the RTL level are much larger than those at the lower levels. Since the RTL power estimates are used to compare several RTL implementation alternatives with a large spread in power dissipation values, a limited accuracy is sufficient.
- Two main approaches are used in power estimation: static [18] and dynamic [5]. Since correlation (both temporal and spatial) can have a large impact on the switching activities [14], only the dynamic (or simulation) approach is feasible for power estimation. Therefore, high-level power estimation should be based on logical simulations, while monitoring the switching activities.
- The parasitic resistance and capacitance of the interconnections are dominating the IC performances in today's multi metal layer processes. Therefore, the interconnect must be taken into account by the power estimation tool.

- A typical digital design flow is based on VHDL. Therefore it is very important to have a power estimator based on that language. In that way it is possible to use the same VHDL design and test-bench descriptions for logical simulation, synthesis, and power analysis.

4.1.1.2 Power Estimation Approaches. Power estimation at a high level has been addressed by several research groups. Basically there are four approaches.

1. **Power estimation based on power models of high-level building blocks**
   Such an approach is presented in [14]. This approach requires a library of pre-characterized and parameterized building blocks, such as ALUs, multipliers, memories, registers, adders, and controllers. This library contains a power model for each of the building blocks, derived from many layout exercises. These models relate the power dissipation to the switching activities of the block-inputs, and to some block specific parameters, such as signal widths, area, and number of states. The switching activities of the block inputs and outputs are obtained from RTL simulations. This approach deals with DSP architectures based on fixed building blocks, and cannot cope with "flexible" architectural synthesis tools [16] or handwritten VHDL code. Each change in technology, synthesis or layout tool, or in the structure of the building blocks requires a considerable re-characterization of the power models.

2. **Power estimation based on *rules of thumb***
   An example is the approach as currently being developed by [19]. This approach estimates the circuit area, and an average value for the switching activity of all nodes, based on the entropy of the inputs and outputs of the circuit. Since no structural information is used it can only give rough power estimations. This approach requires additional fine-tuning before becoming feasible.

3. **Power estimation based on *fast* synthesis**
   Here, an RTL description is synthesized and the gate-level result is analyzed using netlist-level power estimation techniques [4, 21]. This approach suffers from long iteration loops and is restricted to synthesizable descriptions only.

4. **Power estimation based on the semantics of high-level language primitives**
   This approach allows for 'arbitrary' input descriptions. We will elaborate on this approach in the coming sections.

4.1.2 Language based power estimation

The approach is based on simulating a high-level VHDL description while counting the transitions at the inputs and outputs of some VHDL operators. These operators are defined as power primitives, since they are responsible for all power dissipation in the circuit. A library containing the energy dissipation per input and output transition of all power primitives is used to transform these transition counts into power estimates. In fact, this library replaces the library of fixed building blocks [14] described in the previous section. However, the big difference is that the information in our library can be found in the standard cell library, and does not require any power model extractions.

```
p: process(A,B,C,D)
begin
    X <= A+B;
    Y <= C-D;
end process;
```

Figure 4.2 VHDL description of a combinatorial process

Definition:. A power primitive is a VHDL language construct that is mapped by a logic synthesis tool onto hardware, and that dissipates power when there are transitions on its input(s) and/or output(s).

Power primitives fall into two classes, explicit and implicit power primitives. Explicit power primitives correspond to the VHDL operators '(n)and', '(n)or', 'xor', 'not', '+' (adder), and '-' (subtracter). Implicit power primitives have to be inferred from the VHDL description and its semantics. The two implicit power primitives are 'mux' (2-1 multiplexer), and 'ff' (flip-flop) .

Explicit power primitives can be found as function or procedure calls in the VHDL description while implicit power primitives must be recognized from their simulation behaviour.

Consider the VHDL description shown in Figure 4.2, using two VHDL operators (explicit power primitives): '+' and '-'. This description relates the behaviour of outputs X and Y to that of inputs A, B, C, and D. Assume a transition occurs on input A, followed later by a second on input C. The first transition causes a transition on output X; the second a transition on output Y. In other words, the first (second) transition causes the adder (subtracter) to dissipate some energy, respectively. During VHDL simulation the process is activated when an event occurs on one of the four inputs. In this particular case, the '+' and '-' operators are each executed twice, whereas each contributes only once to the power dissipation. Therefore, a method is required to detect if the execution of an explicit power primitive contributes to the power dissipation.

As stated, the implicit flip-flop power primitives should be inferred from the VHDL language. There are two different conditions for a logic synthesis tool to generate memory: (i) incomplete assignments, and (ii) using values from the history (previous values) of signals or variables. These situations have to be detected and inferred from the VHDL description.

We also have to infer multiplexers. For simplicity, consider a 2-1 multiplexer. There are many different ways to describe this multiplexer in VHDL 4.3. However, the hardware implementation of this multiplexer is always the same: a block with two data inputs, one select input and one output. The behaviour of this multiplexer is as follows: If there are no transitions on the select input, then the output will simply follow one of both data inputs. However, each time there is a transition on the select input, the output

```
X <= A(sel);
```

```
Y <= Q when (sel=0) else R;
```

Figure 4.3 Two VHDL descriptions that lead to multiplexers

will switch from one data input to the other. This can cause additional transitions on the output. This switching behaviour is used to detect multiplexers during VHDL simulation.

Another way of inferring multiplexer is a multiple assignment to a variable or signal. If a variable or signal is assigned multiple values (e.g. in different conditional branches), a multiplexer should be inferred.

We refer to [20] for more details on the power primitives.

4.1.2.1 Interconnect Modeling. The number of transitions at the inputs and outputs of the power primitives is one part of the information required to estimate the power dissipation. The other part is the amount of capacitance switched by these transitions.

Consider a netlist, and let G be a gate in this netlist. The total capacitance (C_{tot}) switched by a transition on the output of G is given by the following equation:

$$C_{tot} = C_{out,G} + C_{wire} + \sum_i C_{in,i}$$

Here, $C_{out,G}$ stands for the internal capacitance of gate G connected to its output, (C_{wire} for the interconnect capacitance driven by gate G, and $C_{in,i}$ for the input capacitance of gate i. Furthermore, i encompasses all gates driven by gate G. The interconnect capacitance C_{wire} can be modeled by:

$$C_{wire} = A + B \times FO$$

Here, A and B stand for parameters representing the load independent and load dependent part of the model, respectively. FO stands for the number of gates (fanout) driven by gate G.

Combining both equations yields:

$$C_{tot} = (C_{out,G} + A) + \sum_i (C_{in,i} + B)$$

Note that the total capacitance can be divided into several contributions: one for the driving gate, and a contribution for each driven gate. These contributions are used together with the input and output transition counts of all power primitives, to estimate the power dissipation. This approach is explained based on the VHDL example shown in Figure 4.4. Let us assume there is a transition on signal X. The 'and' power primitive dissipates an amount of energy equal to $(C_{out,AND} + A) \times V_{dd}^2/2$. The 'not' and 'or' operators are executed exactly one Δ unit of simulation time later. Since both operators have one switching input, an amount of energy equal to $(C_{in,NOT}+B)\times V_{dd}^2/2$ is dissipated by the 'not' power primitive, and an amount equal to $(C_{in,OR} + B) \times$

```
X <= A and B;
Y <= not X;
Z <= C or X;
```

Figure 4.4 Three concurrent assignments in VHDL

$V_{dd}^2/2$ is dissipated by the 'or' power primitive. Note that the energy dissipation related to the transition of signal X is broken up in several parts, which are spread out over time. This example illustrates that it is not necessary to explicitly know the fanout of a power primitive, since the VHDL simulator executes the fanout operators automatically at a later time.

4.1.3 Benchmarking the estimation approach

Our approach is based on simulating the RTL or behavioural VHDL description, while monitoring the switching activities of the inputs and outputs of the power primitives. A VHDL power package has been defined containing new VHDL data types and operators, equal to those normally used for RTL design, but extended to store additional power information. A converter has been developed, which reads the VHDL descriptions of the design and test bench, and outputs slightly modified copies based on the types and operators defined in the power package. The modified descriptions are simulated by a VHDL simulator, while counting the number of transitions on the inputs and outputs of the power primitives. After simulation, these counts are converted into power dissipation values, using a table containing the amount of switched capacitance per input and output transition for all power primitives. This capacitance information has been extracted from the power data of the standard cells in a half micron technology.

Complex operators like multipliers and carry- look-ahead adders are dealt with by using executable VHDL descriptions of these operators. These descriptions mimic the hardware behaviour during simulation.

Finally, to get an impression of the overall accuracy, we have analyzed a large number of circuits, such as adders ('Add'), subtracters ('Sub'), multipliers ('Mult'), pipelined multipliers ('Mult pipe'), and IIR filters of order 2 ('IIR'), for both CRA and CLA implementations (Figure 4.5). All input vectors were randomly chosen. The average of both gate level power estimates (synthesis for cost and timing) is plotted on the horizontal axis, and the RTL power estimates on the vertical axis. The average deviation is 40%. The least squares method has been used to determine the best linear approximation to these points. This approximation (slope 1.06) is shown as solid line in the figure. The dashed lines represent the case of a deviation of a factor of two from the approximation.

There is a good correlation between the RTL and gate level power estimates, since the slope of the linear approximation is very close to unity, and since most points are located between the dashed lines.

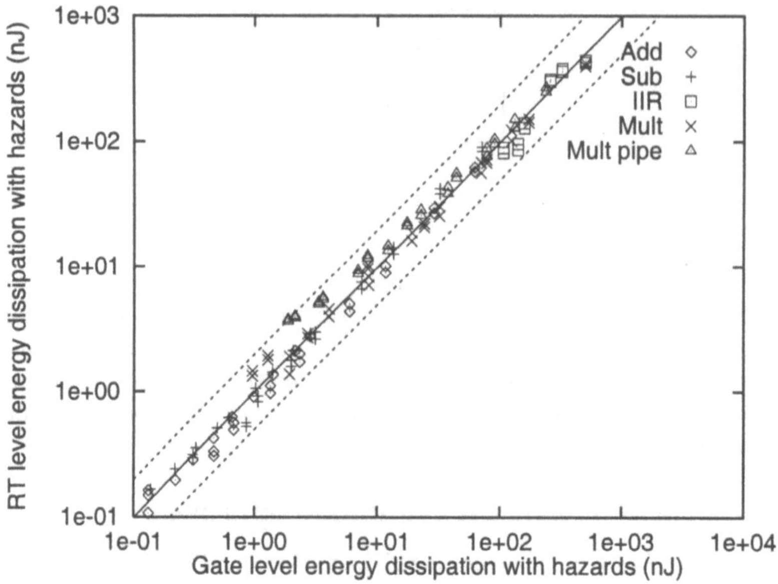

Figure 4.5 Overall accuracy.

4.1.4 Summary

A power estimation approach has been developed to estimate the power dissipation during RTL and behavioural VHDL simulations. The presented approach is flexible, since it is able to deal with any RTL VHDL code and a lot of behavioural code, and it avoids maintaining a library of power models, in contrast to previous approaches. An implementation has been made, based on a VHDL converter, a VHDL package, and power information available in normal standard cell libraries. This has been implemented in the tool PETROL. The presented benchmark demonstrate the flexibility and feasibility of our approach.

4.2 POWER ESTIMATION CASE: DAB SYNCHRO

We have applied the power estimation approach of the previous sections to the DTSE versions of Chapter 3 for the synchronization processing of the Digital Audio Broadcast (DAB) demodulator and channel decoder.

Figure 3.6 in chapter 3 shows the complete DAB chain. Part of the shaded area is subject of this design case. This part (called synchronization processing) takes care of the synchronization between the receiver and the transmitter. It was designed using the architectural synthesis tool MISTRAL2. In order to use the MISTRAL2 tools in combination with DTSE optimizations, we had to modify the standard MISTRAL2 design flow. This section describes the modifications to the MISTRAL2 design flow and the interaction with DTSE approach when applied to our demonstrator application.

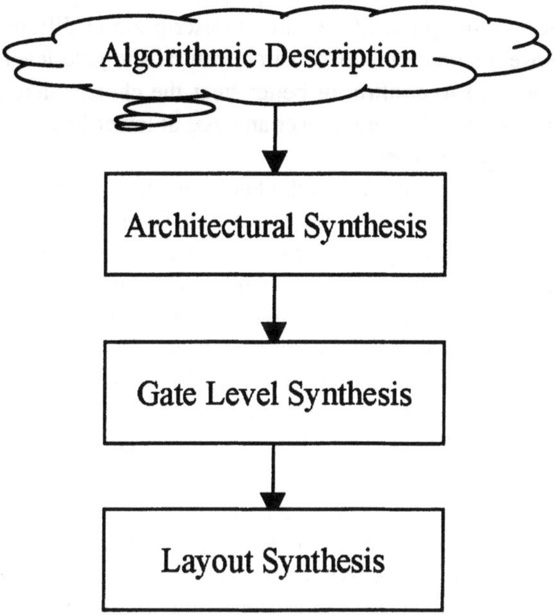

Figure 4.6 Overview of a design flow for Digital VLSI circuits

4.2.1 Design flow

The design flow for VLSI circuits consists of a pile of synthesis steps. There are mainly three synthesis steps involved in the design flow for DAB. Each of them describes the functionality of the integrated circuit at a different level of abstraction. These steps are architectural (or high-level) synthesis, gate-level synthesis and layout synthesis (see Figure 4.6). Each synthesis step generates a description of the circuit starting from a more abstract one. The most abstract description of the task(s) to be implemented constitutes the input for the architectural synthesis step. The input is a program written in a C-like high-level language called DFL (Data-Flow Language), specifying the algorithm's functionality.

Designing a VLSI circuit starting from an algorithmic description offers several advantages. In fact at this high level of abstraction it is easy to widely explore the design space in order to meet specific constraints on area, cycle budget, cycle time and power dissipated. The quick feedback given by architectural synthesis allows the designer to try out different design solutions while keeping the final time-to-market low [17].

In this high-level synthesis context, the methodology for low-power design as used in the DAB project is situated. The low-power methodology (ATOMIUM-ADOPT) manipulates the program code of the algorithm to be implemented so that the synthesis flow generates an ASIC with reduced power consumption [8, 7]. In order to have an early feedback on the effects of the low power methodology, it is very useful to estimate the power consumption of the circuit description at different levels of abstraction. Early power estimation between architectural and gate-level synthesis steps for instance, can be used to confirm or better steer the choices taken in the synthesis flow itself. This can avoid longer power analysis at lower levels of abstraction on ineffective processor architectures.

Thus, the design and the analysis methodology for low power are strongly combined and integrated in a single flow. Power analysis, far from being just a step in the overall design flow, can be thought as a process whose results are refined while going from the specification to the physical implementation of the IC.

Figure 4.7 shows a schematic description of the overall flow we use to demonstrate the benefits of the ATOMIUM low-power methodology for the DAB example. The program code for the synchronization processing of the DAB channel decoder, here considered, is written in DFL. This program constitutes the input for the architectural synthesis tool MISTRAL2. The original DFL program is manipulated and transformed to achieve a low power implementation using the ATOMIUM-ADOPT methodology. The output of the high-level synthesis is an RTL-VHDL description of a processor specifically tuned for the given application. An early estimation of power dissipation is performed at this point with the PETROL high-level power estimation tool as presented in Section 4.1. During the same simulation, the activity of the processor is traced and the operations performed by the functional units in the processor are monitored. The effectiveness of the low power methodology and the architectural synthesis choices are validated comparing the power analysis results with those obtained from the original design. If the low-power goals are not reached, then a new iteration in the design flow is performed. This means either perform different transformations on the original program during the ATOMIUM low-power methodology, or modify the architectural synthesis choices taken, or even reviewing both the steps. A gate level synthesis is then run using one of the commercially available synthesis tools. A final and more refined power analysis is performed at the gate level with the gate-level power estimation tool DIESEL. DIESEL is an in-house simulation-based power estimation tool that simulates a gate-level netlist and while simulating counts the transitions of the signals. Together with a power model of the standard cells and a wire-load model, the dissipation is calculated.

4.2.2 Power Analysis: General Criteria

The aim of the analysis methodology is to prove and to verify the effectiveness of the low-power design flow. Hence, the analysis flow should hold the following characteristics:

- **Flexibility**
  The source code, though implementing the same algorithm, can be far different from the original, after the transformations induced by the low-power optimization. Running the architectural synthesis tool with the new source code as input for the synthesis while keeping the synthesis directives defined for the original source code will generate an inefficient processor in terms of latency, datapath

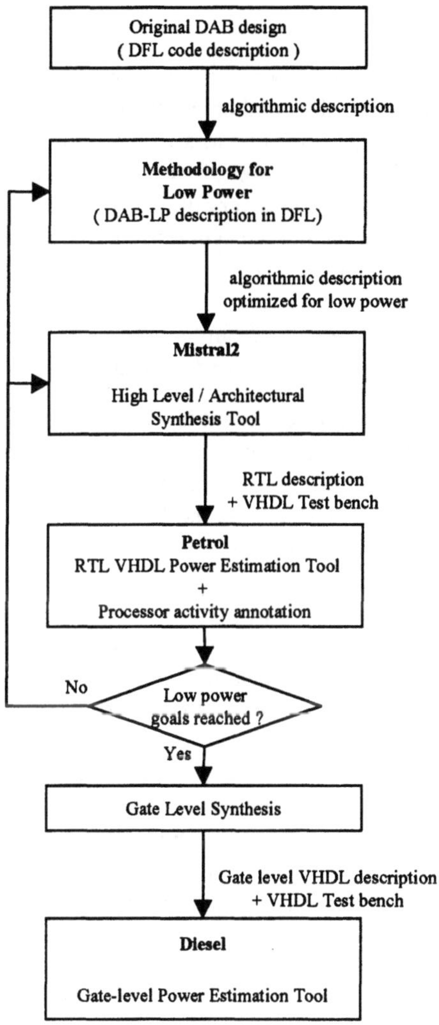

Figure 4.7 DAB overall design flow

resources, control area, power. Clearly, this can spoil the benefits provided by ATOMIUM low-power methodology. Therefore, the architectural synthesis step has to be guided and tuned for every single program provided for low power, in order to generate good results. The processor generated by MISTRAL2 will differ for each DAB program given, therefore the subsequent power analysis has be flexible as well.

- **Highly automated**
  The interaction between the power analysis results and the low-power methodology itself can modify the number and the type of transformations applied on the original code. This brings us to a new loop in the flow where again new code is synthesized and the power of its implementation is estimated. Running through this design flow loop is very time consuming and a large effort must be invested to achieve a high quality synthesized processor for every new program code given. Hence, a methodology is needed to obtain good synthesized architectures in a systematic way for a set of low-power programs. Moreover, we need a methodology to extract from the processors generated by the synthesis step the power results in a correct and automated way.

- **Provide power analysis at different level of abstraction**
  The power analysis is performed in the DAB-LP project at two different levels of abstraction. The PETROL tool estimates the power of the processor described at a RTL-behavioural VHDL while the DIESEL tool performs the analysis using a gate level description for the ASIP. The time needed to perform a power estimation at the gate level depends on two factors: the time needed to have a gate-level synthesis of the whole processor with one of the available tools (e.g. Synopsys), and the time needed and often underestimated to write or modify parts of the VHDL source to get a fully synthesizable description. The strength of the first power analysis performed by PETROL is that power can be estimated even if the VHDL description of the architecture is not yet entirely synthesizable. Hence, an early feedback can be obtained at this level. Thus, the second power analysis should be performed only when a sufficiently deep design space exploration has been carried out and both the high-level program DFL code and the synthesized processor are fairly optimized. The results with the second analysis are intended to verify and confirm the quality of the whole low-power design flow so far.

The next sections cover a number of issues mentioned above. The first section copes with the problem of generating a high-quality processor by means of architectural synthesis, starting from the DAB low-power modified program. A novel mapping step is proposed to reduce the number of resources in the datapath while keeping the performance, the functionality and the benefits for power consumption introduced by ATOMIUM methodology. The tool (SFGNURSE), which improves the mapping step and the overall synthesized design, is here described. The interaction between SFGNURSE and the MISTRAL2 environment is shown as well. A new architecture, stemming from the standard MISTRAL2 one, is introduced. The extended address computation functionalities of the modified architecture are illustrated as well. The

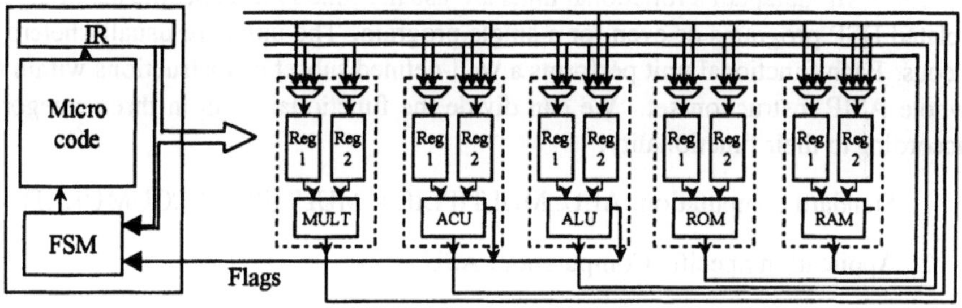

Figure 4.8 MISTRAL2 architecture template

second section deals with the power analysis approach. Here the method to trace the processor's activity is explained. Finally, the language based power estimation is used to compare the original design and the optimized design.

4.2.3 Architectural Synthesis Modifications

4.2.3.1 Mistral2 Algorithm Mapping step. The target architecture of the MISTRAL2 high-level synthesis tool is a DSP processor. The instruction-set implemented is specifically tailored towards the program to be executed. For this reason it is commonly referred to as an ASIP (Application Specific Instruction-set Processors). The ASIP generated, belongs to the class of Harvard processors, hence with a separate memory for data and instructions. In order to enhance parallelism a Very long Instruction Word (VLIW) architecture has been chosen. This means that several execution units in the datapath can be independently driven by the control to perform an operation during the same cycle.

The MISTRAL2 architecture consists of a datapath and a control unit. The control unit consists of a microcode memory, where the instructions for the datapath are stored, and a FSM used to compute the next instruction address based on the current state and on the flags coming from the datapath. That datapath consists of many execution units connected to one another via buses. Each Execution Unit (EXU) is made up with a core, a number of register files and multiplexers (see Figure 4.8).

The core is actually the element that performs the operation. Each core input is connected to a register array (register file) where temporary data used by core is stored. At the top of each register file lays a multiplexer whose inputs are connected to datapath buses. The aim of the mux's is twofold i.e. they allow the core to process data produced by EXUs and they bit-align the values coming from other units in a way suitable for the core.

The mechanism of moving data from one functional unit to another is commonly referred in literature as "data routing". Another way to accomplish the same results is "spill code generation", namely to use a background memory to exchange data from one unit to another. This method is usually exploited when the registers in the architecture are a "scarce resource".

The ASIP datapath's functional units are usually tuned for a specific set of closely related DSP programs or even for a single program. The units are usually heterogeneous. Each functional unit performs a well-defined subset of instructions within the whole ASIP instruction set. We can divide the functional units in three categories according to their functionality:

1. Standard computation: ALU, MULTIPLIER, MULTIPLY-ACCUMULATE
2. Application Specific Computation: ASU
3. Address Computation: ACU

The Address Computation Unit (ACU) is a unit specifically designed to perform operations needed for addresses, such us modulo addition and subtraction, bit-reversal operations, decrement&test. The instruction set of an ACU partially overlaps with the ALU instruction set.

One of the most critical MISTRAL2 steps for a good synthesis of the DAB code, is actually the mapping step. Here the algorithm is translated to a number of Register Transfer (RT) instructions. Moreover, here the functions in the code are mapped to real resources. The Algorithm mapping step works on a Signal Flow Graph (SFG), which is actually a different representation of thc original program. In the SFG the functions are represented by nodes, and the dependencies between functions and data by edges (arrows). The goal of the step is to map the program on the resources given in the architecture. To accomplish this task, the mapping procedure needs a description of the architecture. This description is given in a template file, which contains a list of all the EXUs available to build up a processor. For each of those EXUs the whole instruction set is listed. The template file is therefore an exhaustive description of all the resources and the instructions available to map a given program on a MISTRAL2 architecture.

The mapping step starts by inspecting the SFG. For each node in the graph the template file is scanned looking for a match. The first match found is then taken to map the considered function to an operation in a functional unit.

The manipulations applied to the code for low-power are mainly related to array accesses and index computations. MISTRAL2 performs the translation from index computation to physical memory address computation during the algorithm mapping step as well. Beside the pattern-based mechanism, some more rules are applied whenever address expressions have to be mapped. The most important rule related to address expressions is the following:

- *A loop counter is processed on an ACU by default*

Given the following array's index expression:

$$C_0 + C_1 * iter_1 + \ldots + C_n * iter_n + other\_expr$$

Where:

- $C_0, C_1, \ldots, C_n$ are integer constant values

- $iter_1, \ldots, iter_n$ are iterators (loop-counters) for addressing elements in the array
- $other\_expr$ is the remaining part of the expression, which is neither a constant nor a product between a constant and an iterator.

The $C_i * iter_i$ terms are mapped on the ACU, where the multiplication are replaced by a sequence of additions in order to limit the complexity of address calculation (strength reduction). The additions among the $C_i * iter_i$ terms are mapped on the ACU as well. The rest of the expression is usually mapped on other functional units according to the template file.

Here are some examples:

`A[5+2*I+J]`

The `2*I` is mapped as an addition → `I+I`. The address calculation is here computed with 4 additions.

`A[I+J*K]`

Because J and K are both variable the "*" is here mapped on a multiplier according to the template file. The result of the multiplication and the iterator I are added on the ACU.

`A[I+J>>2]`

In this case, the ">>" is mapped on an ALU while the addition is still mapped on ACU because the ACU is not able to perform shift operation.

4.2.3.2 Why a further architectural optimization.

The low-power methodology proposed in the DAB project, aims to minimize the number of accesses to data-memories of big dimensions while enhancing the usage of little memories or even better of registers in the datapath (Chapter 3). This power consumption reduction is obtained by modifying the DFL code of the algorithm to be implemented. As a consequence of the transformations, data locality is more exploited and the energy spent to access data from memories is greatly decremented. Nevertheless, this can produce drawbacks and negative effects on other parts of the synthesized processor, for instance on the microcode memory.

Before showing some possible artifacts generated by an ineffective synthesis, some considerations must be stated:

- *The width and the type of data values and address values are usually different.* In ASIP designs, fixed point is preferred to floating point arithmetic in order to achieve the best performance with the minimum amount of computational area. So width and alignment of data-types is strongly dependent on the algorithm to be implemented and on the accuracy needed on the output data. On the other hand, address-types are usually unsigned integers and the width of the address type is usually smaller compared to the data type and is related to the memory requirement needed to run the program.
- *Complex computations are performed moving data from one functional unit to another through buses and mux's.* As a consequence of this, the architectural synthesis tool, if not properly guided,

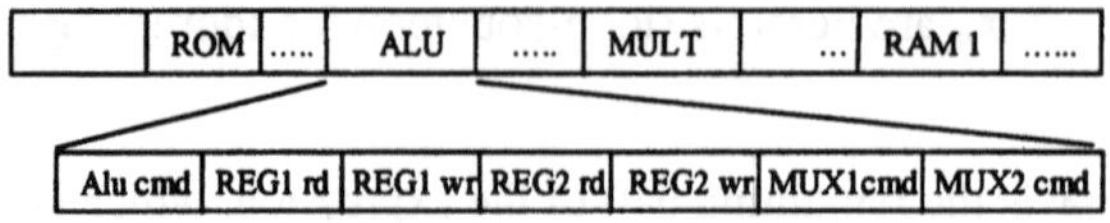

Figure 4.9 Fields and subfields in a Very Long Instruction Word

will generate an inadequate architecture in terms of area, and latency. The benefits of a code optimized with respect to the number of memory accesses might be, in this way, lost. The major drawbacks are:

- a longer schedule
- more multiplexers to route data from one unit to another to perform address computations
- more registers because of data copies spread out in the register files of the architecture.

The above mentioned drawbacks will indirectly influence the size of the microcode memory. In fact, it will be longer because of the longer schedule. And it will be larger, because now more bits are necessary for the control unit to drive the register files and the multiplexers.

The instruction word (see Figure 4.9), which carries the commands from the control unit to the datapath's EXUs, determines the width of the microcode memory itself. A command in the instruction word is composed of several subfields, each of them addressing a component within an EXU. From the last figure it is clear that the width of the total instruction is highly dominated by the width of those subfields needed to drive registers and multiplexers. Thus, minimizing the number of register per register files and the number of mux inputs is fundamental to keep the width of the overall instruction word small. In this way the power dissipated in the microcode is reduced.

The methodology used to tune the high-level synthesis for the set of low-power programs is the following:

- *Split address computation from real data computation.*
  This means that every computation related to addresses will be performed by the ACU and every data operation by the remaining units (see Figure 4.10). Because data-related variables and address-related variables will never be processed by the same core, there is no need for data alignment or routing from ACU to ALU, anymore. Thus, less data aliases will be spread in different registers and fewer connections between EXUs will be generated. This will reduce the overhead of mux's and registers in the architecture drastically .

- *Extend the instruction set for the ACU.*
  Because data and address computation has been split, now the ACU must be

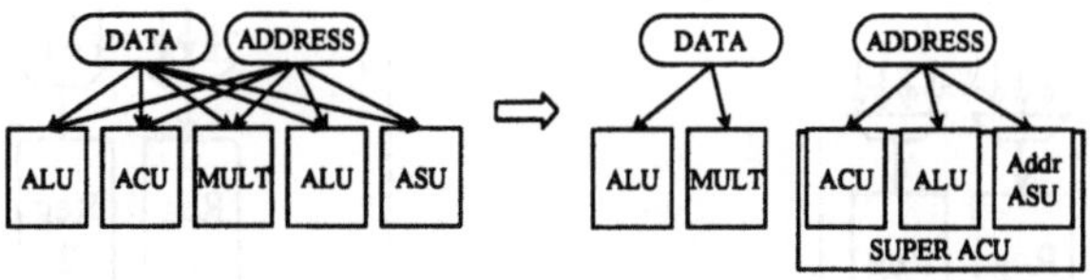

Figure 4.10 Splitting DATA from ADDRESS computation

SuperACU instruction set	
ACU instruction set	
add, sub, inc, dec addmod, submod, incmod, decmod bitrev	shiftup, shiftdown Multiply by a constant and, or, not (bitwise) $<, >, ==, !=, <=, =>$

Table 4.1 ACU and SuperACU instruction sets

Flags	Purpose
GT, LT, GE, LE, EQ, NE	for comparisons between the first and the second operand of the SuperACU
GTZ, LTZ, GEZ, LEZ, EQZ, NEZ	for comparisons between the first operand and the value zero.

Table 4.2 Flags generated by the SuperACU

capable of executing every possible instruction needed for address computation. The ACU instruction set must be extended with new operations. This new EXU with a superset of instructions is here called SuperACU (see Table 4.1). Moreover the new proposed SuperACU can generate the flags of Table 4.2 for the control unit.

Figure 4.11 shows the internal architectures for a standard ACU and for the proposed SuperACU. The input data coming from the shared register files are distributed to different sub-cores. The sub-cores employed to cover the instruction set of the SuperACU are: SACU-acu, SACU-alu, SACU-asu. The command for the EXU core is internally sliced in 3 sub-commands each of them going to a different sub-core. Only one sub-core can be active per clock cycle. The command bus is also used to drive the output multiplexer. In this way the output of the active sub-core is then selected to be the output of the SuperACU. The SuperACU produces also a larger number of flags. The set of flags is the union of the flag set generated by each sub-core.

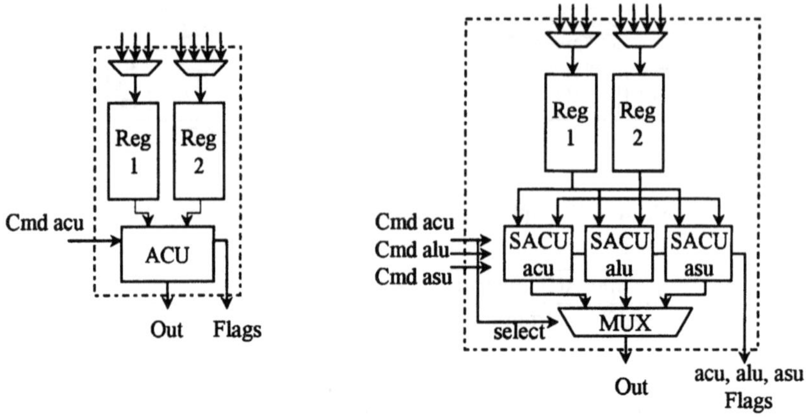

Figure 4.11 ACU and SuperACU internal architectures

Extending the instruction set does not necessarily mean having more hardware than that needed with the standard ACU. In fact if some of the instructions in the SuperACU instruction set are not used for a given program, they can easily be stripped out, together with the hardware implementing them. That means that also the area of the SuperACU can be reduced.

The SuperACU offers the possibility to analyze the effects on the design of the low power techniques applied on the DAB program. The SuperACU architecture is the element needed to enhance the flexibility of the power analysis, it will help understanding how important the address computation issue in the overall low power methodology is.

4.2.3.3 Address Operation Mapping. Consider the approach used to generate address computation during the algorithm mapping step. The main characteristic of the whole approach is a great simplicity and transparency. Nevertheless such an innocent-looking mapping criterion can easily hide low-quality mapping as shown by the following two examples:
Example 1:

```
J = I+2;
K= 1+2*J;
A[K] = ...
```

Suppose that the loop-counter or iterator is I. Because the mapping method works really locally it will not be able to map correctly the functions needed to calculate K on an ACU correctly even though the address calculation performed is not different from the one needed for A[5+2*J].
Example 2:

```
A[1 +3*(2*J-2)+2*(K-J-4)]
```

Scanning Direction ↓

Add, sub, rsub, inc, ...	Arithmetic	ALU
In1 and In2 comparisons In1 and zero comparisons	Logic	
Add, sub, inc, ...	Arithmetic	ACU
Modadd, modsub, modinc, ...	Modulo	
==0, ==1 (designed for loop bound check)	Logic	
Multiply		MULT
Mult-acc positive Mult-acc negative		MULT-ACC
Application Specific		ASU

Table 4.3 Template structure and scanning direction

As described in section 4.2.3.1, the algorithm mapping step needs a well determined pattern to map address expressions. Hence, the expression of example 2 will be only partially matched. Clearly it could be mapped entirely on an ACU, but due to the local scope of the mapping technique, which actually can be considered as a parsing step, the two expression 3*(2*J-2) and 2*(K-J-4) will be mapped as "general expression" and therefore mapped on a common ALU.

The mapping step starts inspecting the SFG. For each node in the graph the template file is scanned looking for a match. The first match found is then taken to map the considered function to an operation in a functional unit. Because of the sequential nature of the scanning process the order how the patterns for matching are written in the template file is very important. The general rule is to give precedence to data processing, and then to address processing. Table 4.3 shows the mapping order given in the file. This order gives precedence to data mapping and exploits the ACU only for the additional features like modulo operations. Because ACU and ALU instruction sets overlap, priority is given to ALU mapping. Now, looking from the address-mapping point of view this scanning order between ACU and ALU needs to be exactly reversed! Clearly a fixed order in the template file and a local scope for mapping function to EXU instructions will never achieve good results both for data and address mapping. The tool available for architectural synthesis leaves these considerations to the designer and gives him the power to exactly map the address computation to the correct unit (ACU) by means of directives to the synthesis tool called *pragmas*.

The list of pragmas necessary to guide the mapping tool toward an efficient address mapping has to be reviewed or even rewritten for every new DFL program. This problem is determined by the fact that variables in the DFL code are looked as meaningless *patterns*. On the other hand, to identify all address-related variables, the real meaning of every variable in the overall computation must be analyzed. Thus, a variable or an operation is address-related if it contributes to an address computation.

The computation which is needed to calculate a certain index can be very complex and spread out over different parts of the code. The power of looking to the semantic

```
func  main    ( Re, Im : fix<32>[1024] ;              // Data In
                enhance : int<8>;                     // Parameter for Data
                a0, b0, range : unsigned_int<10> )    // Parameter for Addressing
                Re_out : fix<32>[146] =               // Data Out
begin
var temp : fix<32>;
var start, stop : unsigned_int<10>;

...

block_#x :: begin

       loop1::(cnt : a0 .. (b0 + range + ref*96 + 64) ) ::begin

              temp := abs ( Re[cnt+1]*Im[cnt+1] - Re[cnt]*Im[cnt] );

              if temp > 0.9
                     then  temp := temp >> 2 ;
                     elsif temp < 0.4 then temp := temp * enhance;
              endif;

       start:= Max( cnt-a0)%32 , cnt-a0-(ref+1)*96+32 );
       stop := Min( (cnt-a0), range);

              loop2::( iter : start .. stop) ::begin
                     rot := (( cnt -a0 - iter )>> 5)%3;
                     re_temp := re_temp
                     if rot == 0 then Re_out[iter] := Re_out[iter] + temp;
                     endif;
                     if rot == 1 then Re_out[iter] := Re_out[iter] - temp;
                     endif;
              end; // end of loop2 , inner loop //
       end; // end of loop1, outer loop //
end; // end of the block_#x ///

...
end; // end of the program //
```

Figure 4.12 DFL example for data-address splitting

of functions in the code is that a correct mapping is guaranteed, no matter how "far" an address-related function is from an array indexing operation.

These considerations are further clarified by an example (see Figure 4.12) The fragment of code shown in the figure is possible low-power code generated after ATOMIUM manipulations. The variables and the operations involved in address computation are in bold characters while data computation and the variables linked to data computation are in normal text. A mapping step based on pattern matching and without a global view is not able to map all the bold variables as address related and therefore some of those will be inevitably mapped on ALUs.

Table 4.4 shows the variables in the code and how they are related to data or address computation. It shows exactly what "dividing address and data computation" means. For example the variable "rot" depends on variables which are all used to calculate the index "iter" for "re_out" array, namely "cnt", "a0" and "iter" itself. Therefore "rot" is semantically an address-related variable. Looking at the signal graph flow, splitting address from data computation, means actually to partition the graph in two subgraphs according to the above mentioned semantic criterion [1, 13, 9].

Address variables:	Data variables:
address inputs: a0, b0, ref, range	data inputs: Re[ ], enhance
iter = f1 (start, stop) start = f2 (cnt, a0, ref) stop = f3 (cnt, a0, range) cnt = f4 (a0, b0, ref, range) rot = f5 (cnt, a0, iter)	temp = g1(Re[ ], Im[ ], enhance) Re_out[ ] = g2(Re_out[ ], temp)
Operation performed for address computation :	Operation performed for data computation
"+", "-", "*96", "inc", "dec", "eqz", ">>", "% (mod.)"	"+", "-", ">", "<", "*"

Table 4.4 Data and Addresses semantically separated

4.2.3.4 SfgNurse tool. The SFGNURSE tool deals with the problem of separating data and address computation. The input of the tool is the SFG graph. The SFGNURSE can be described in an algorithmic fashion as follows:

```
SfgNurse ( SFG )
{
   For each fetch or store operation:
      Identify the index variable;
      Tag&Propagate ( variable );

   Inspect the graph:
      The tagged variables are address-related variables;
      The tagged operations are address-related operations;

   Write a PRAGMA file, steering all the tagged objects
    to be computed by the SuperACU.
}

Tag&Propagate (variable)
{
   Tag the variable;
   Find in the SFG all the edges connected to that variable;
   { Find and tag all the operations
     where the variable is consumed or produced;  }
     For each operation :
         For each input or output variable in the operation :
            Tag&Propagate ( variable);
}
```

The SFGNURSE's core is based on a recursive procedure, namely tag&propagate() which traverses the code, throughout all the hierarchy levels in the graph, looking for dependencies among variables.

Figure 4.13 shows a small example how the tool actually works. The address re-

```
#define ADDR unsigned_int<10>
#define DATA int<24>

func main(  var  in   : DATA[8] ;
                 c    : DATA;
                 mode: unsigned_int<2>)

            var  out  : DATA[8] =
begin
var a : ADDR;

   a:= 4 + mode;

   loop_1::(cnt:a  .. 7)::  out[7 - cnt]   := in[cnt] - c*2;

end;
```

Figure 4.13 Simple DFL code fragment to show how the SFGNURSE works

lated variables, constants and operations are written in bold characters. There are two different index variables in the code. The first, explicitly defined, is "cnt", the second is implicit defined and correspond to "7-cnt". Those two index variables are related to the variables "a" and "mode". From this program code it is also evident that some constant values are used only for address computation ("7", "4", '8") while other only for data computation ("2"). Figure 4.14 shows the equivalent SFG of the given code.

Arrays and array index are involved in two kinds of operations:

1. Fetch operation: this operation has two inputs, the array and the index variable. The output of the operation is the element of the array at the index position.

2. Store operation: the operation has two inputs the element to be inserted in the array and the array index. The output is the array updated with the new element at the index position.

The tool starts looking for a fetch or a store operation. Suppose that the fetch instruction is found first. The input connected to the index variable is taken (in the picture marked with a black spot), and the variable 'cnt' is tagged. Then all the operations connected to the variable 'cnt' are tagged, namely "-", "inc", ">" (shadowed in the figure 4.14). At this point the algorithm keeps on finding address-related objects applying the same procedure recursively to the variables connected to the just found nodes. The analysis propagates also throughout the hierarchy. In this way the "+" operation, which produces the variable "a" in the main block, is tagged as well. This algorithm is repeated for each fetch or store operation present in the SFG. This guarantees that all the address-related operations are exhaustively tagged.

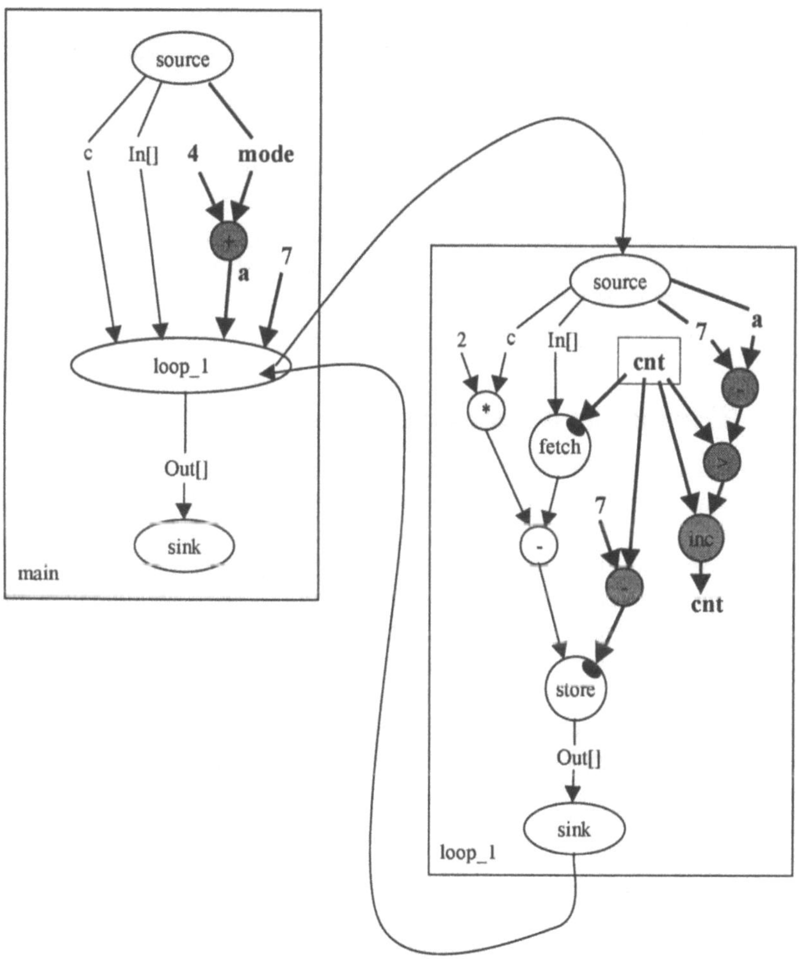

Figure 4.14 Hierarchical SFG for the DFL code in Figure 4.13

4.2.3.5 The Mistral2 flow revisited. In order to have a high quality architectural synthesis, several modifications have been introduced in the original Mistral2 design flow (Figure 4.15).

- A preprocessing step to automatically generate pragmas for ACU computation (SFGNURSE).
- A new, flexible ACU with an extended instruction set (SuperACU).
- The template file has been updated.
  New patterns for the algorithm mapping step have been added in the file to support the SuperACU functionality.
- The library file has been updated.
  This file contains the bit-patterns associated for each EXU's opcode. The bit-coded representation of the instruction is needed to fill the microcode memory with the correct sequence of ones and zeros. The correspondence between instruction opcodes and bit-patterns is given in the Mistral2.lib file. A longer list of bit-patterns is given for the SuperACU, because now the unit is capable of executing more instruction.

4.2.4 Tracing the processor activity

The modifications applied to the original DAB program have also effects on the control graph and on the hierarchical structure of the SFG. The loops in the low-power program and in the subsequent signal flow graph are data-dependent. Hence, it is not possible to determine the overall latency of the processor just looking at the number of microcoded instructions generated by Mistral2 for the ASIP. Furthermore, the activity of the ASIP cannot be derived directly from the scheduled sequence of VLIW instructions, again because of data-dependent loops and because of conditional blocks which are executed only when some condition on some variable are true.

In order to trace the activity, the RTL-VHDL description of the ASIP is simulated using a meaningful input stimuli file. The activity both in the datapath and in the controller is dumped in a file and some statistics and parameters are derived from the file itself.

For each building block in the ASIP architecture (register, controller logic, mux, cores, memories), a synthesizable VHDL description exists. For each VHDL entity description, a new VHDL "architecture" has been written which provides the same functionality but also some monitoring functionality.

Here is an example showing how a behavioural VHDL model of a MUX is monitored:

```
LIBRARY ieee;
USE ieee.std_logic_1164.ALL;
USE ieee.std_logic_arith.ALL;

ENTITY mux IS
   GENERIC(width     : INTEGER := 8;
```

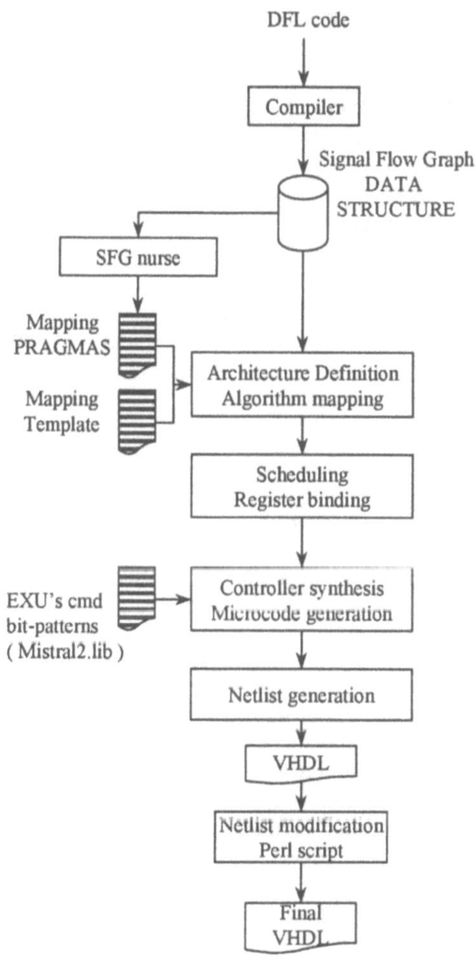

Figure 4.15 Mistral2 architectural synthesis flow revisited

```
          log_ports : INTEGER := 2;
          ports     : INTEGER := 4);
   PORT  (xbus      : IN  std_logic_vector((width) * ports - 1 DOWNTO 0);
          mxbus     : IN  std_logic_vector(log_ports - 1 DOWNTO 0);
          obus      : OUT std_logic_vector(width - 1 DOWNTO 0));
END mux;

ARCHITECTURE monitor OF mux IS

BEGIN
  mux : PROCESS(xbus, mxbus)
    VARIABLE output: std_logic_vector(width - 1 DOWNTO 0);
  BEGIN
    FOR i IN 0 TO ports - 1 LOOP
      IF is_x(mxbus) THEN
        output := (OTHERS => 'X');
      ELSE
        IF i = to_integer(mxbus) THEN
          output := xbus((i * width) + width - 1 DOWNTO i * width));
      END IF;
    END LOOP;
    obus <= output;
  END PROCESS;

  MUX_tracing : PROCESS(mxbus)
  BEGIN
       mux_monitor(mux'Instance_Name, ports, mxbus);
  END PROCESS;

END;
```

As shown in the VHDL code a procedure to monitor activity information is called whenever an event on "mxbus" happens. In order to monitor different instances of a "mux" entity in the overall ASIP architecture an VHDL '93 attribute is used, namely 'Instance_Name. This allows tracing independently the activity of different mux's in the architecture. This is the procedure written for monitoring a mux.

```
LIBRARY ieee;
USE ieee.std_logic_1164.ALL;
USE ieee.std_logic_arith.ALL;
USE ieee.std_logic_textio.ALL;
USE std.textio.ALL;

PACKAGE trace_pck IS
   CONSTANT dump_file : string := "monitor.out";
   FILE f : TEXT IS OUT "monitor.out";
   PROCEDURE mux_monitor(mux_instance : IN string ;
                         max_ports    : IN integer;
                         port_number  : IN std_logic_vector);
END trace_pck;
```

```
PACKAGE BODY trace_pck IS
 PROCEDURE mux_monitor (mux_instance : IN string ;
                        max_ports    : IN integer;
                        port_number  : IN std_logic_vector) IS
   VARIABLE l : line;
   BEGIN
      IF NOT(is_x(port_number)) THEN
        write(l, NOW, Right, 13, ns);
        write(l, mux_instance & string'(" : "), Right, 15);
        write(l, string'("["));
        write(l, max_ports );
        write(l, string'("] , port n. "));
        write(l, to_integer(port_number));
        writeline(f, l);
      END IF;
   END mux_monitor;
END trace_pck;
```

The monitoring mechanism is quite flexible. A given instance of a unit can be monitored or not, just changing the name of the architecture in the configuration file. A typical dump file, with the monitoring results after the simulation looks like:

```
  12.5 NS  mreg_3_ram_1   :  read(     0)=0000000000
  12.5 NS  mreg_3_ram_2   :  read(     0)=0000000000
  37.5 NS  acu_1_alu_1    :
  37.5 NS  acu_1_acu_1    :  add
 237.5 NS  mreg_3_ram_1   :  read(   200)=0000000000
 237.5 NS  mreg_3_ram_2   :  read(   200)=0000000000
 312.5 NS  acu_1_acu_1    :  decx
 337.5 NS  acu_1_acu_1    :  sub
 362.5 NS  acu_1_acu_1    :  incx
 387.5 NS  acu_1_acu_1    :  sub
 387.5 NS  mreg_3_ram_1   :  read(   199)=0000000000
 387.5 NS  mreg_3_ram_2   :  read(   199)=0000000000
 437.5 NS  acu_1_acu_1    :  add
 462.5 NS  acu_1_acu_1    :  sub
 537.5 NS  acu_1_acu_1    :  add
 562.5 NS  acu_1_alu_1    :  aslx
 587.5 NS  acu_1_alu_1    :
 587.5 NS  acu_1_acu_1    :  sub
 612.5 NS  acu_1_alu_1    :  sub
 612.5 NS  acu_1_acu_1    :  add
 662.5 NS  acu_1_alu_1    :
 662.5 NS  acu_1_acu_1    :  sub
 687.5 NS  acu_1_acu_1    :  passx
 712.5 NS  acu_1_acu_1    :  sub
 737.5 NS  acu_1_acu_1    :  add
 762.5 NS  acu_1_alu_1    :  passx
 787.5 NS  acu_1_alu_1    :
```

The dump file can be filtered (post-processed) in order to get the required information, e.g. the number of operations executed on the SuperACU, the number of register accesses, etc. The exact information that should be extracted is very much depending on the phase were the design is in, and should be selected on an ad-hoc basis. However the dump file contains all the information.

4.2.5 RT-level power analysis results

In this section we compare the power dissipation of the original DAB synchronization processor design with the power consumption after DTSE optimizations. In order to do this, the design was re-designed using the modified architectural synthesis flow as described in Section 4.2.3.

The PETROL tool of Section 4.1 is used to analyze the power based on simulation of the RTL code generated by MISTRAL2. The same analysis method is applied to both the original design (also called reference design) and the DTSE-optimized design (also called optimized design).

Figure 4.16 shows the different distribution of power consumption of different blocks for both designs. It shows that the code-ROM power consumption is dominant (the code-ROM is called "rom" in the figure). Furthermore, because of the longer schedule, the low-power design dissipates some more power for its code ROM. Together with the longer schedule, the optimized design introduces some extra computations. This eventually increases the power spent in the processor (datapath and controller). To remedy this, significant changes in the microcode processor architecture are needed. In particular, instruction cache improvements in terms of both architecture and compiler support have been addressed (see [3] and their references). They are especially based on adding a small loop cache [2] or "filter cache" [12], between the L1 instruction cache and the instruction decoding. These work especially well if the instruction cache and decoder are distributed and placed very locally to the parallel FUs in the processor [6]. Once this problem is tackled, we have focus on the second most important issue, namely the data transfers and storage contributions in the register-files (foreground memory), the SRAM and the DRAM.

In the data-related large memories (in particular in the DRAMs) the power is heavily reduced. Once the instruction-set decoding bottleneck is removed, the total power consumption of the whole synchro core can be significantly reduced further. Figure 4.17 shows the effects of the DTSE methodology (without applying the ADOPT stage yet) on the datapath and on the controller. One of the effects of the methodology in the DAB code is that data is reused without repeatedly accessing large memories. This means also that more computations are executed in the datapath. In particular the ACU is more extensively used due to the more complex control flow in the optimized code. This leads also to more complex address expressions compared to those present in the reference design. Next, the ALU also solves some extra conditions, especially to determine whether array bounds are still respected. These effects can be minimized by properly applying the ADOPT techniques.

Re-using data also implies an increment of power dissipated in foreground register files in the processor's VLIW datapath. The extra power dissipated in the registers is caused by two factors. First, more registers are needed in the architecture to store

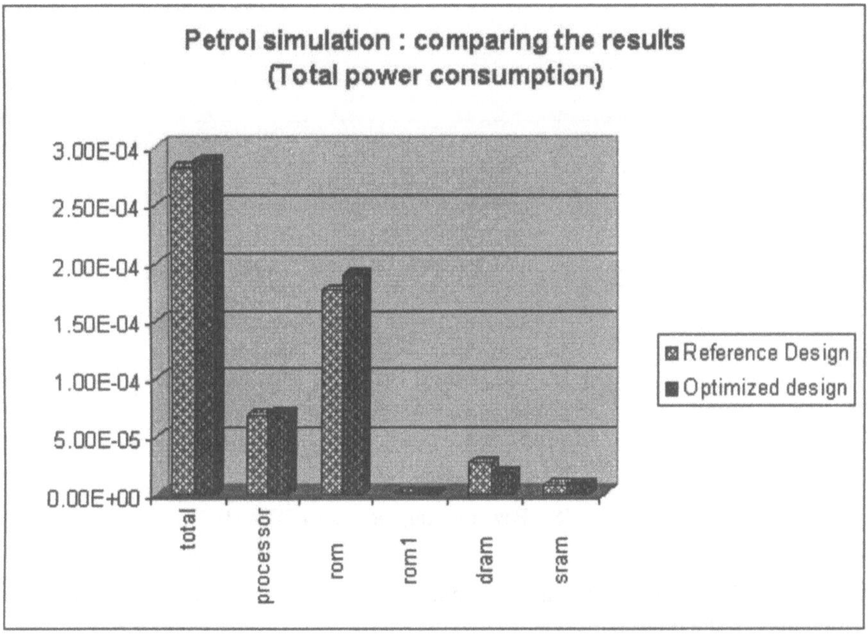

Figure 4.16 Power comparison: reference and optimized DAB synchro core

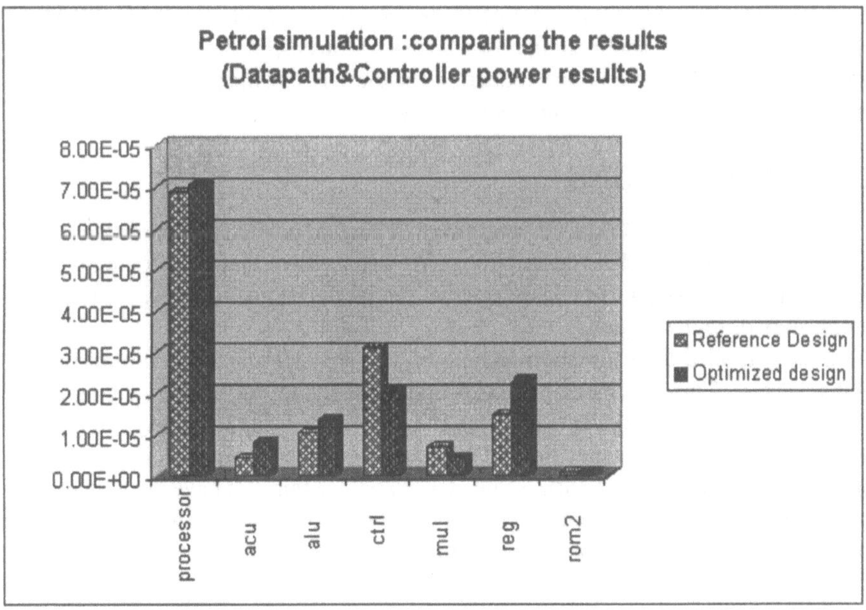

Figure 4.17 Reference and optimized DAB synchro datapath&controller (no memories)

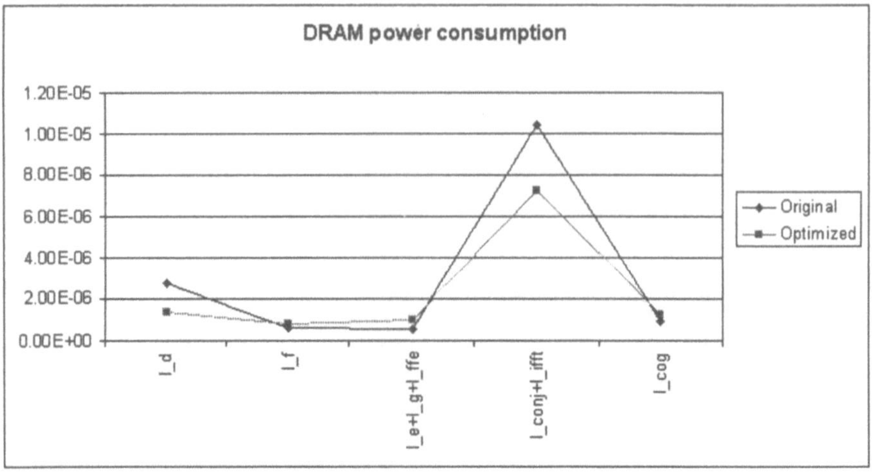

Figure 4.18 Power savings accessing DRAMs

values kept for further computations. Secondly, registers are more often read and written than in the reference code.

The ATOMIUM methodology produces its best results in the power reduction of data-related memories. Figures 4.18 and 4.19 show the power consumption as a function of the phase of the synchronization algorithm in the DRAMs and SRAMs respectively. The DTSE methodology of Chapter 3 tries to reduce accesses to large memories (DRAMs) possibly at the cost of more accesses to small memories (SRAMs). This results in a small increment in power for SRAM accesses. However, this is counterbalanced by a dramatic power reduction for the DRAM memory accesses. Note that the power scale of the two figures is different.

4.3 CONCLUSIONS

In this chapter we have introduced the concept of language-based power estimation for RTL and behavioural VHDL descriptions. We have shown that the accuracy of this approach is high enough to make high-level design decisions with respect to power optimization of digital circuits.

We have applied the estimation techniques to the synchronization processor of a DAB receiver. This processor has been optimized using the DTSE techniques of Chapter 3. We have analyzed the power consumption for both the original and the optimized design. For this specific VLIW processor, power dissipation in the instruction memory is the dominant part. Some ideas to reduce this part have been identified. The DTSE techniques have clearly reduced the power consumption in the data related storage and traffic, which is the second contributor in the original power budget.

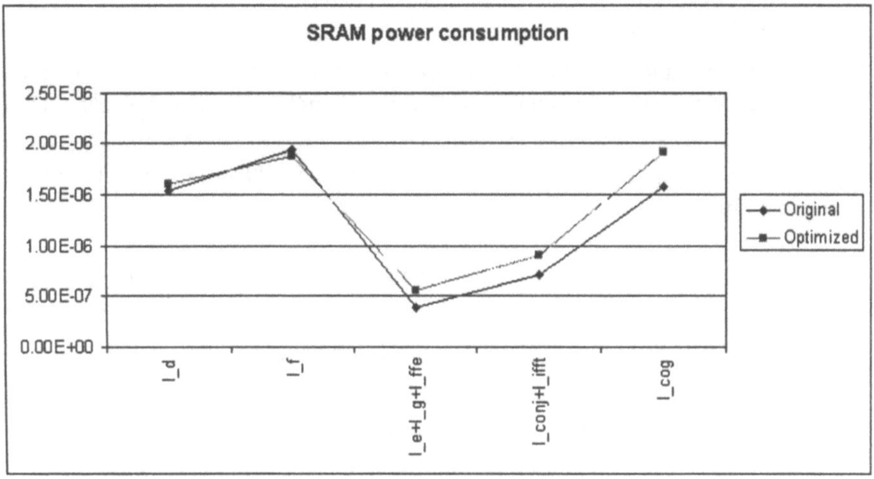

Figure 4.19 Power results: accessing SRAMs

References

[1] A. Aho, R.Sethi, and J. Ullman, "Compilers: Principles, Techniques and Tools", *Addison-Wesley*, 1988.

[2] R.Bajwa, M,Hiraki, H.Kojima, D.Gorny, K.Nitta, A.Shridhar, K.Seki, K.Sasaki, "Instruction buffering to reduce power in processors for signal processing", *IEEE Trans. on VLSI Systems*, Vol.5, No.4, pp.417-424, Dec. 1997.

[3] N.Bellas, I.Hajj, C.Polychronopoulos, G.Stamoulis, "Architectural and compiler support for energy reduction in the memory hierarchy of high-performance microprocessors", *Proc. IEEE Intnl. Symp. on Low Power Design*, Monterey CA, pp.70-75, Aug. 1998.

[4] A. Bogliolo et al., "Adaptive Least Mean Square Behavioral Power Modeling", *Proc. ED&TC*, 1997, pp. 404–10.

[5] R. Burch et al., "A Monte Carlo Approach for Power Estimation", *IEEE Trans. on VLSI*, Mar. 1993, pp. 63–71.

[6] F.Catthoor, "Energy-delay efficient data storage and transfer architectures: circuit technology versus design methodology solutions", invited paper, *Proc. 1st ACM/IEEE Design and Test in Europe Conf.*, Paris, France, pp.709-714, Feb. 1998.

[7] F.Catthoor, F. Franssen, S.Wuytack, L. Nachtergaele, and H. De Man, "Global communication and memory optimizing transformations for low power signal processing systems", *IEEE workshop on VLSI signal processing*, Oct. 1994.

[8] F.Catthoor and M. Cupak. "Efficient functional validation of system-level loop transformations for multi-media applications", *Proc. Electronic Circuits and Systems Conference*, Bratislava, Slovakia, Sep. 1997, pp 39–43.

[9] N. Deo. "Graph Theory with applications to engineering and Computer Science", *Prentice-Hall*, 1974.

[10] European Broadcast Union, "Radio Broadcast Systems; Digital Audio Broadcasting to Mobile Portable and Fixed Receivers", *prETS 200401*, Vol. Final Draft 1994.

[11] Le Floch, B., Halbert-Lassalle, R., and Castelain, D., "Digital sound broadcasting to mobile receivers", *IEEE Transactions on Consumer Electronics 35*, 3, Aug. 1989.

[12] J.Kin, M.Gupta, W.Mangione-Smith, "The filter cache: an energy efficient memory structure", *Proc. 30th Int. Symp. on Microarchitecture*, pp.184-193, Dec. 1997.

[13] D. Kuck, "The Structure of Computers and Computation", *Wiley*, 1978.

[14] P.E. Landman and J.M. Rabaey, "Architectural Power Analysis: The Dual Bit Type Method", *IEEE Trans. on VLSI*, Jun. 1995, pp. 173–87.

[15] D. Liu and C. Svensson, "Power Consumption in CMOS VLSI Chips", *IEEE J. of Solid State Circ.*, Jun. 1994, pp. 663–70.

[16] J.L. v. Meerbergen et al., "PHIDEO: High-Level Synthesis for High Througput Applications", *J. of VLSI Signal Processing*, Vol. 9, 1995, pp. 89–104.

[17] G. De Micheli, "Synthesis and optimization of digital circuits", *McGraw Hill*, 1994.

[18] F.N. Najm, "Transition Density: A New Measure of Activity in Digital Circuits", *IEEE Trans. on CAD*, Feb. 1993, pp. 310–23.

[19] M. Nemani and F.N. Najm, "Towards a High-Level Power Estimation Capability", *IEEE Trans. on CAD*, Jun. 1996, pp. 588–98.

[20] Peset Llopis, R., "A New Approach in High-Level Power Estimation", *Proc. DATE*, 1998.

[21] Power Compiler of Synopsys, 1996

[22] H.J.M. Veendrick, "Short-Circuit Dissipation of Static CMOS Circuitry and its Impact on the Design of Buffer Circuits", *IEEE J. of Solid-State Circ.*, Aug. 1984, pp. 468–73

[23] "IEEE Standard VHDL Language Reference Manual". *IEEE Std 1076*, 1993.

5 LOW-POWER CUSTOM REGULAR PROCESSOR SYNTHESIS FLOW

Roger Woods, Gayle Lightbody, Jonathan Spanier, Gareth Keane

DSP and Telecommunications Group, Queen's University of Belfast

r.woods@ee.qub.ac.uk

In high-throughput signal processing applications that are both arithmetic- and data-dominated, the performance capabilities of customized processor solutions are attractive particularly in applications such as video compression where functionality is fixed or has been standardized. For high volumes, customized VLSI solutions offer the ultimate system performance in terms of both area and speed. For example, a processing performance of 100 GOP/s/cm^2/W is achievable for computationally complex DSP algorithms in 0.35μm standard cell CMOS technology. The performance gain is possible as it is often possible to exploit characteristics of the algorithm in such a way to allow a highly optimized implementation to be realized. A number of different design flows exist for this target domain. In this chapter, we will focus on relatively regular signal processing kernels that are critical in the overall performance and cost of the global application. Examples include motion estimation, discrete orthogonal transforms, algebraic solvers, Fourier transforms and the like. The designs resulting from our tuned approach are characterized by high area utilization, high levels of locality (preserving power) and efficient memory utilization.

Broadly speaking, the development of DSP processor solutions can be classified into two design styles.

- The first approach involves the development of a instruction set processor where suitable processing units are synthesized to achieve the required performance. Once a sensible architecture has been determined such as the processor in the previous chapters, the majority of the design activity is involved in scheduling, memory optimization and code transformations. As the underlying architecture is well understood and highly predictable, accurate performance in terms of

F. Catthoor (ed.),
Unified Low-power Design Flow for Data-dominated Multi-media and Telecom Applications, 97-134.

area, speed and, of course, power, is relatively well understood. Techniques for reducing power in these structures have been addressed by several researchers.

- The second approach which is described in this chapter does not use a prescribed architecture but attempts to synthesize an application specific architecture. The close relationship between the architecture and the function or functions that it implements, allows the designer to apply a number of performance efficient optimizations that would not be available in the previous design flow. In addition, the designer has access to hardware transformations and different number systems which can offer advantages in some applications. The system-level data transfer and storage aspects for the main (background) signals in such kernels can benefit from the techniques described in the chapter 3 but the foreground memories and the custom processor core of these architectures requires additional techniques.

The material described in this chapter concentrates on the synthesis of the custom processor core for the second classification. This latter approach allows a more power-efficient solution as the architecture can be customized to meet an area, speed and power budget. On the other hand, this flexibility presents greater challenges in delivering low power solutions as there are a greater number of degrees of freedom. In addition, the work attempts to address the development of the customized regular processor in a parameterized fashion as this is a key area of research in the field of Intellectual Property (IP) cores. This involves developing a VLSI architecture that can be applied across a wide range of applications. For example, consider the applications of the Discrete Cosine Transform (DCT) that are listed in table 5.1. This outlines how key parameters such as point size and throughput rate vary across a number of applications including image compression and speech coding. For an effective implementation, the designer needs to develop a solution that has a single architecture that is defined in terms of parameters such as point size, input and output data sizes and coefficient wordlengths. This process requires the development of an architectural solution that is scalable and has a simple control interface. This solution can then be coded as a parameterizable VHDL description allowing the design to be highly portable and to be easily reused. The process of developing a parameterized solution is described in this chapter through the development of a DCT implementation. In addition, the general process of developing a regular processor is illustrated in the development of a Recursive Least Squares filtering circuit for adaptive beamforming.

5.1 LOW-POWER REGULAR PROCESSORS

Chandrakasan [Chandrakasan et al., 1992] defined four levels at which power consumption can be tackled. These are algorithm, architectural, circuit and technology. Techniques at the algorithmic level include investigation of word length determination and data representation as presented in [Chandrakasan and Brodersen, 1996], development of low power algorithms, voltage minimization techniques for processors and system level code transformations. In our unified flow (see chapter 2), these are more refined and they correspond to the algorithm synthesis, data type refinement and the task- and data-level stages. At the architectural level, the use of concurrent

Table 5.1 Practical DCT Applications

Application	*1D DCT*	*2D DCT*	*Operating Speed*
Image Compression	...	4x4 8x8 16x16	5 Mbps (slow motion video) to 1.33 Gbps (HDTV)
Filtering	8 12 16 32 64 128	8x8 16x16	30 Mbps (Progressive image transmission)
Transmultiplexing	3 4 7 14 64 72 128	...	64kps (speech)
Classification	...	8x8	30 Mbps (video file transfer)
Vector Quantization	4 9 16 25	2x2 5x5 6x6	250Mbps
Speech Coding	32 64 128 256 512	...	64 kbps (speech)

processing techniques [Chandrakasan et al., 1995] and locality [Mehra et al., 1996] has resulted in considerable reductions in power consumption. These correspond to the instruction/operation level stages, especially the processor architecture integration, the custom processor synthesis and the instruction-set processor mapping substages. At the circuit level, corresponding to our technology mapping stage, alternative circuit organizations such as Complementary Pass Logic (CPL) [Zimmermann and Fichtner, 1997] and other design styles such as asynchronous circuits [Vanhoof et al., 1993] have been used. At the (process) technology level, low power technologies [Foty and Nowak, 1994] such as Silicon on Insulator (SOI) technology [Stork, 1995] have been considered. Variable level voltages such as those described in [Kuroda et al., 1998] have also been explored.

Generally speaking, power reduction techniques either act to minimize the switched capacitance ($C * F$) or trade off performance with reduced supply voltage [Chandrakasan et al., 1992]. In voltage minimization techniques, transformations are used to speed up the system's throughput beyond that which is necessary. The voltage is then reduced, slowing up performance until the required throughput rate is met but at a lower power consumption budget. However, in the semi-custom approach used in many design flows, the voltage of the technology is pre-determined by the silicon foundry and cannot be altered by the user to reduce power. Typically, the voltage will have already been developed for low power operation and the user will access the design though library phantoms which the user cannot modify. Therefore, the "reduced voltage" techniques cannot be employed.

Low power design therefore must be targeted at reducing the switched capacitance. This is the summation of the products of the toggling occurring on each node in the circuit by the capacitance of that node. It is important to consider switched capacitance as opposed to toggling or circuit capacitance independently. For example, a circuit may have a large capacitative net which has a low switching activity and which will not contribute greatly to power consumption. Conversely, low capacitance nets may have a lot of switching activity. This feature is demonstrated in the detailed study of multiplier architectures outlined in this chapter where some of the multipliers have lower power but have a similar degree of toggling. A subsequent detailed analysis reveals the source of the power consumption. For this reason, the concept of minimizing switched capacitance in order to reduce power consumption is adopted for the regular processor meta flow.

5.2 CUSTOM PROCESSOR DESIGN FLOW FOR REGULAR SIGNAL PROCESSING MODULES

The major focus in this chapter is to develop an approach to the design of regular VLSI processors. These are processors where the architecture of the circuit is developed to meet the specific requirements of a computationally complex part of the overall processor. The emphasis is to develop low power versions that map to the meta flow path trajectory as given in chapter 2. The starting point of the regular processor meta design flow is an algorithmic description of the functionality that has to be mapped to the processor. This algorithmic description is then refined using the design flow illustrated in figure 5.1, resulting finally in a circuit level solution.

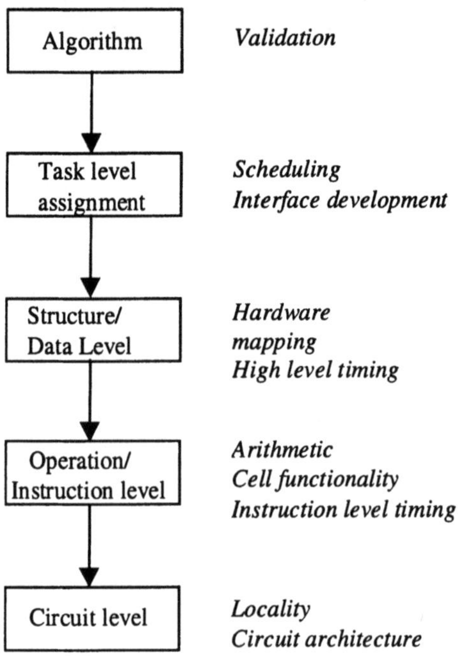

Figure 5.1 Regular processor design flow for signal processing modules

algorithmic: This involves the generation of a correct algorithmic description for the specific custom processor e.g. a full MPEG encoder. At this stage, the main focus is to develop and validate the algorithmic operation. High level modeling tools such as Cadence SPW^{TM} and Matlab^{TM} can be used to assist in this purpose. This stage will be addressed in section 5.4.

data type refinement: At the algorithm stage, the development will have been carried out using abstract data types, such as floating point arithmetic. In DSP design, a fixed point or a limited range floating point solution is preferable as it usually results in a more hardware efficient solution. In many cases, the choice of fixed or floating point arithmetic is not straightforward and can have a dramatic effect on the system performance. This will include the consideration of finite wordlength issues such as truncation and saturation. These issues will not be elaborated in this chapter.

task level: The algorithm is broken down into a number of separate tasks. For an MPEG encoder, the various tasks will include the DCT and motion estimation. Typically, the final solution will be implemented as a full custom processor or the computationally complex parts of the system will be implemented as a custom processor. A key issue is to ensure that the minimal number of processors are used in the task allocation process. In a parameterizable IP core context, the task allocation will have been pre-determined in the design process for the core. So also this stage is not further discussed in this chapter.

parallel-array/data level issues: At this stage, the data parallel functions will be transformed into an underlying structure of processors. For largely regular DSP algorithms, this will involve the development of a VLSI architecture from a Dependence Graph (DG) representation. A considerable body of work exists into the transformation of commonly used functions such as the DCT and FFT into efficient structures where the number of multiplications and additions are reduced. A key issue will be an efficient use of hardware for the given application requirements. This stage will be addressed in section 5.5.

operation/instruction level issues: A number of choices are available at this stage including the structure of the processor blocks, arithmetic representation and data control. For example, the multipliers and adders can be implemented as separate processors or they may be mapped together in a multiply-accumulator structure (In regular array multiplier structures, the cost of modifying the array to permit accumulation has a small overhead). The power consumption can vary with the different bit level architectures. The choice of data type is highly important as the regular-processor designer has access to a wide choice of options e.g. fixed point and different types of arithmetic e.g. redundant arithmetic and residue arithmetic representations. The choice of the different arithmetic representations i.e. other than two's complement, will be dictated by specific performance advantages gained. For example, redundant arithmetic can be used in the development of recursive structures such as recursive digital filters and in arithmetic functions that are naturally most significant bit first, namely division and square root. This stage is the main focus of this chapter, see section 5.6.

circuit level: The circuit level implementation is beyond the scope of this book but preserving regularity in the generation of the circuit level description and use of pipelining registers in the datapath can result in power savings.

5.3 GENERAL PRINCIPLES

As previously discussed, the key objective for low power operation is to reduce the switched capacitance of the circuit. A number of approaches in order to achieve this goal have been detailed below.

- The development an architecture with localized connections. This acts to minimize the capacitance component of power consumption.
- Application of mappings to ensure high utilization of the processors. This acts to reduce the overall hardware requirements and therefore power.
- Reduction of the overall circuit switching. This can be achieved in several ways including the use of hardware parallelism and redundant arithmetic number representations which eliminates carry propagation chains.
- Movement of registers through the architecture to minimize both the number and size of the data registers. For example, banks of registers may be added to balance datapaths. Using well established rules, these registers can be moved through the architecture to nodes with smaller datapath sizes or large capacitative nets or to locations which help to reduce critical paths.
- Choice of suitable low power arithmetic processors. One example is in the implementation of division. A special purpose divider can be used or the operation can be expressed as a series expansion and mapped onto a multiplier.

5.4 ALGORITHMIC ISSUES

There have been a wide range of techniques for developing a mathematical representation from an algorithmic description and it is beyond the scope of the book to discuss all of these. The process in figure 5.2 shows the general design flow for development of highly regular, VLSI solutions for DSP algorithms. The initial specification is given as a mathematical relationship which is translated into a DG representation. A DG representation shows the data dependencies between different dataflows. At this stage, it is possible to develop a Signal Flow Graph (SFG) representation and a number of projections can be applied to develop SFG alternatives that match the algorithmic requirements. The bit-level processors are then embedded into the SFG. These may have detailed timing requirements which may mean retiming the SFG to ensure correct operation. From a low power perspective, the final VLSI solution should ideally meet some or all of the following criteria.

- Circuit representation with localized connections.
- Solution with efficient wordlength representations.
- Optimized dataflow to minimize transitions on the internal circuit buses.

Another criteria is to develop a solution that is parameterizable. This is a key aspect to the formation of an IP block. The concept of a parameterizable architecture allows

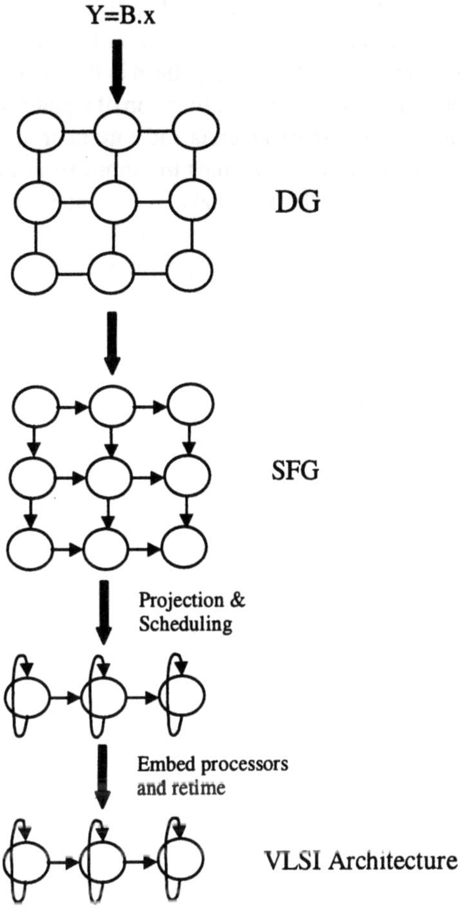

Figure 5.2 Development of VLSI architecture

some level of accurate power estimation to be maintained across a range of parameters. This is important in cases where the design needs to be modified at a latter stage of the design flow. A parameterizable block for Recursive Least Squares filtering is used to demonstrate some of these issues in the chapter.

5.4.1 Recursive Least Squares Filter

In adaptive filtering two conflicting algorithms dominate the field namely, the Recursive Least Squares (RLS) algorithm and the Least Mean Squares (LMS) algorithm. The RLS algorithm is a powerful technique derived from the method of Least Squares. It offers greater convergence rates than its rival LMS algorithm but this gain is at a cost of increased computational complexity, a factor that has hindered its use in real time applications.

One application is in the area of beamforming. The function of a typical adaptive beamformer is to suppress signals from every direction other than the desired "look direction" by introducing deep nulls in the beam pattern in the direction of the interference. The beamformer output is a weighted combination of signals received by a set of spatially separated antennae, one primary antenna and a number of auxiliary antennae (Figure 5.3). The primary signal, y_n, constitutes the input from the main antennae, which has high directivity. The auxiliary signals, X_n, contain samples of interference threatening to swamp the desired signal. The filter eliminates this interference by removing any signals in common with the primary input signal.

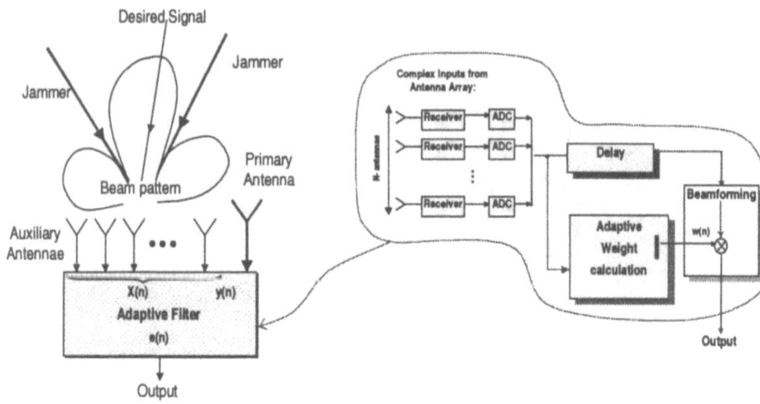

Figure 5.3 Beamforming application

There has been a considerable body of work devoted to algorithms and VLSI architectures for RLS filtering with the aim of reducing this computational complexity such as that by Guorong and Yungong [Guorong and Yungong, 1990] and Radar [Rader, 1996]. Much of this work has concentrated on calculating the inverse of the correlation matrix, required to solve for the weights, in a more stable and less computationally intensive manner than a straightforward matrix inversion. The standard RLS algorithm achieves this by recursively calculating updates for the weights using the matrix inversion lemma [Haykin, 1986]. An alternative and very popular solution performs a set of orthogonal rotations, e.g. Givens rotations [Givens, 1958], on the incoming data matrix transforming it into an equivalent upper triangular matrix. The filter parameters can then be calculated by back substitution. This method, known as QR decomposition, is an extension of QR factorization that enables the matrix to be re-triangularized, when new inputs are present, without the need to perform the triangularization from scratch. From this beginning, a family of numerically stable and robust RLS algorithms has evolved from a range of QR decomposition methods such as Givens rotations and Householder transformations [McWhirter, 1983].

A matrix X(n) is decomposed into an upper triangular matrix R(n) and an orthogonal matrix Q(n) such that:

$$X(n) = Q(n)R(n) \tag{5.1}$$

The X(n) matrix is pre-multiplied by rotation matrices one element at a time. The rotation parameters are calculated so that the sub-diagonal elements of the first column are zeroed. Then the next column's sub-diagonal elements are zeroed and so forth until an equivalent upper triangular matrix is formed. Givens performs this operation through a sequence of rotations, which are best described by a trivial example. Consider a 2x3 matrix,

$$\begin{bmatrix} a_{11} & a_{12} & a_{13} \\ a_{21} & a_{22} & a_{23} \end{bmatrix}$$

This matrix is transformed into a pseudo-triangular matrix by eliminating the element, a_{21} by multiplying the matrix by the rotation matrix,

$$\begin{bmatrix} \cos\alpha & \sin\alpha \\ -\sin\alpha & \cos\alpha \end{bmatrix}$$

Thus

$$\begin{pmatrix} \cos\alpha & \sin\alpha \\ -\sin\alpha & \cos\alpha \end{pmatrix} \begin{pmatrix} a_{11} & a_{12} & a_{13} \\ a_{21} & a_{22} & a_{23} \end{pmatrix} =$$

$$\begin{pmatrix} a_{11}\cos\alpha + a_{21}\sin\alpha & a_{12}\cos\alpha + a_{22}\sin\alpha & a_{13}\cos\alpha + a_{23}\sin\alpha \\ -a_{11}\sin\alpha + a_{21}\cos\alpha & -a_{12}\sin\alpha + a_{22}\cos\alpha & -a_{13}\sin\alpha + a_{23}\cos\alpha \end{pmatrix} \tag{5.2}$$

To eliminate a_{21}, solve for $-a_{11}\sin\alpha + a_{21}\cos\alpha = 0$. Thus,

$$a_{11}\sin\alpha = a_{21}\cos\alpha \tag{5.3}$$

From the trigonometry relationship shown in figure 5.4, the pseudo matrix is achieved using the shown values for $sin\alpha$ and $cos\alpha$.

$$\begin{bmatrix} a_{11new} & a_{12new} & a_{13new} \\ & a_{22new} & a_{23new} \end{bmatrix}$$

This operation can be achieved by a number of different circuit architectures such as the enhanced Squared Givens Rotation algorithm [Walke, 1997]. A SFG for this circuit and cell definitions are shown in figures 5.5 and 5.6 respectively. The next number of sections looks at the derivation of a circuit architecture from the algorithmic description that meets specific requirements.

5.4.2 Derivation of the RLS filter

The specific process of developing a VLSI architecture for the RLS filter from the algorithmic description is shown in figure 5.7. The data dependencies can be identified

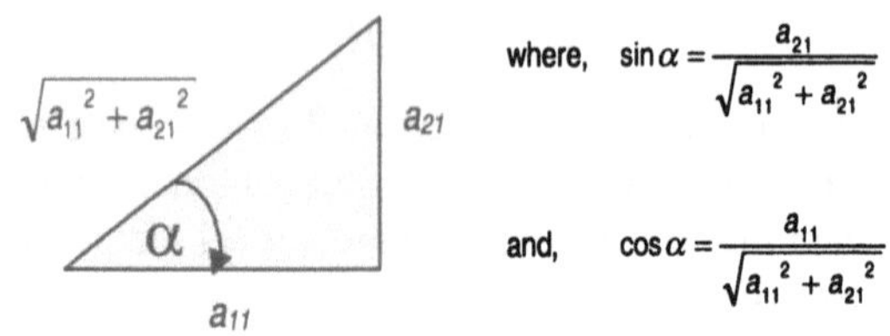

Figure 5.4 Applying the rotation to eliminate a_{21} results in the pseudo-triangular matrix

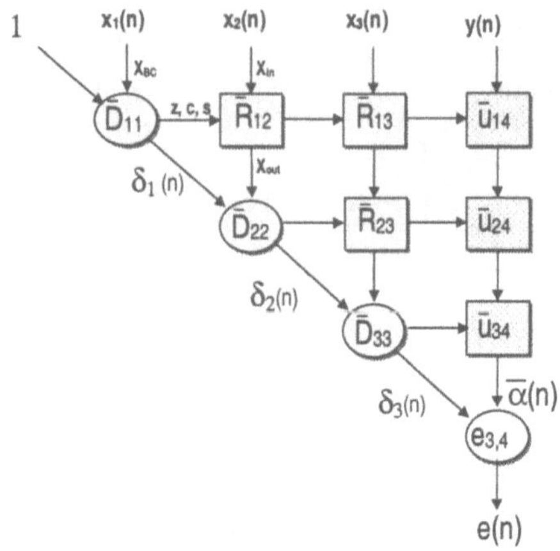

Figure 5.5 QR array for the SGR algorithm

in the DG which allows the maximum level of concurrency to be identified by breaking the algorithm into nodes and arrows. The nodes outline the computations and the direction of the arrows shows the dependence of the operations. This is shown for the QR algorithm by the three dimensional DG in figure 5.8. The diagram shows three successive QR iterations, with arcs connecting the dependent operations. Some of the variable labels and the most fight hand column of cells in figure 5.5 has been omitted for clarity.

In summary, the QR array performs the rotation of the input $\underline{x}^T(n)$ vector with R values held within the memory of the QR cells so that each input x value into the boundary cells are rotated to zero. The same rotation is continued along the line of internal cells via the horizontal arrows between QR cells. From this DG it is possible to derive a number of SFG representations. The most obvious projection, which is used here, is to project the DG along a time axis i.e. in the R direction.

To derive the SFG from the DG, the nodes of the DG are assigned to processors and their operations are scheduled on these processors. One common technique for processor assignment is linear projection of all identical nodes along one straight line

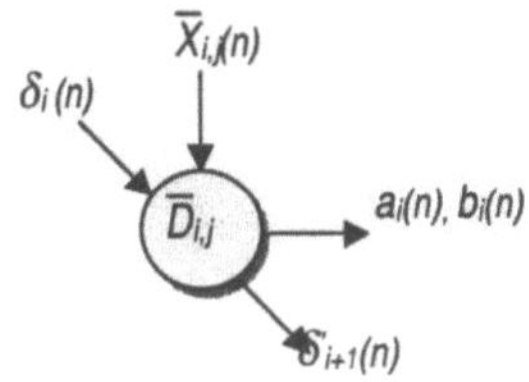

$$a_j(n) = \delta\overline{X}_{i,j}(n)$$

$$b_j(n) = \frac{\overline{X}_{i,j}(n)}{\overline{D}_{i,j}(n-1)}$$

$$\overline{D}_{i,j}(n) = \beta^2\overline{D}_{i,j}(n-1) + \delta\overline{X}_{i,j}{}^2(n)$$

$$\delta_{i+1}(n) = \beta^2\frac{\overline{D}_{i,j}(n-1)}{\overline{D}_{i,j}(n)}\delta_i(n)$$

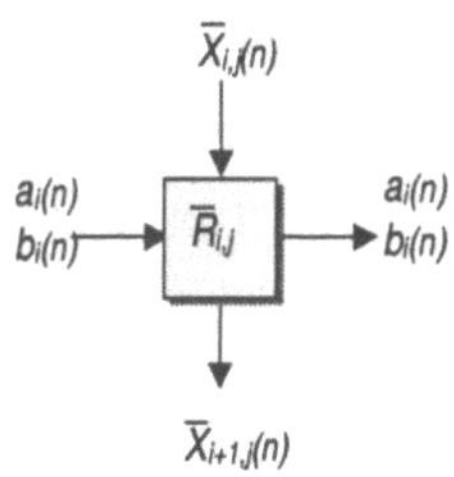

$$\overline{R}_{i,j}(n) = \beta^2\overline{R}_{i,j}(n-1) + a_i(n)\overline{X}_{i,j}(n)$$

$$\overline{X}_{i+1,j}(n) = \overline{X}_{i,j}(n) - b_i\overline{R}_{i,j}(n-1)$$

Figure 5.6 Cell definitions for the QR array

onto individual processors. This is represented mathematically by the projection vector d (Figure 5.9). Linear scheduling is then used to determine the order in which the operations are performed on the processors. The schedule lines in figure 5.9 indicate the operations that are performed in parallel at each cycle. Mathematically they are represented by a schedule vector s that is normal to the schedule lines, and that points in the direction of dependence of the operations. That is, it shows the order in which each line of operations is performed.

There are two basic rules that govern the projection and scheduling, and ensures that sequence of operations is retained. Given a DG and a projection vector d, the schedule is permissible if and only if :

- All the dependence arcs flow in the same direction across the schedule lines.
- The schedule lines are not parallel with the projection vector d.

The SFG graph can be translated into an architecture by physical assignment of operations to hardware modules, and secondly, by scheduling the operations onto these modules. The techniques used to assign the operations to processors is the same as for the SFG derivation from the DG. An architecture may be derived directly from this SFG by allowing a processor for each node in the diagram.

5.5 PARALLEL-ARRAY/DATA LEVEL ISSUES

The previous section suggested a triangular array DG for RLS filtering. If each of the cells are implemented directly, an triangular implementation of a typical size, namely 41, that is clocked at 100MHz, processes data at 800 GFlops. This capability is well

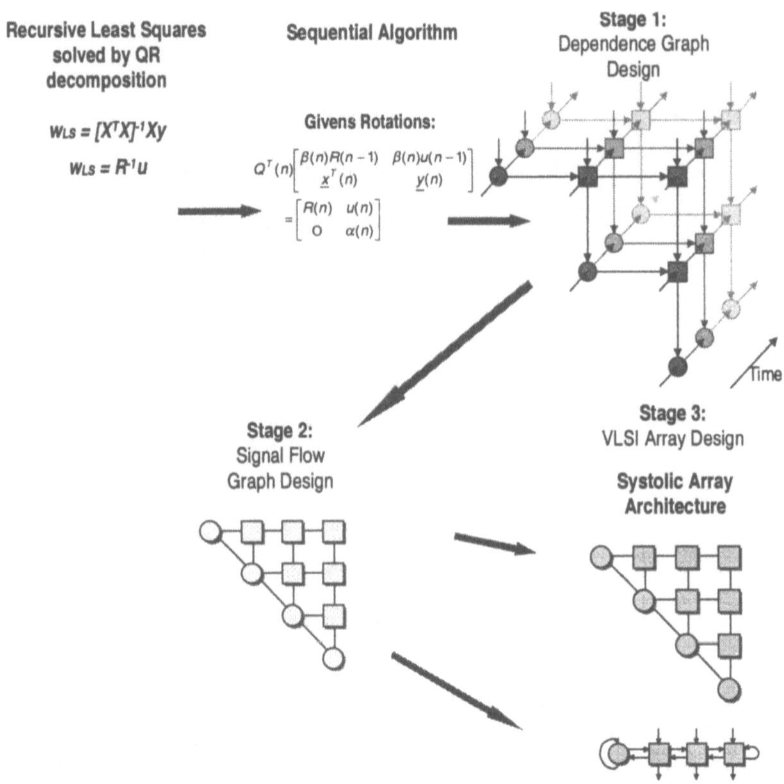

Figure 5.7 General process for developing VLSI architectures

beyond that required for most applications. Thus the goal is to achieve an highly regular implementation with localized connections that achieve 100% utilization.

There are several ways of efficiently mapping the architecture, in this case, a triangular array, down onto a lower number of processors. Two of these are given below.

- The first approach is to work out a detailed scheduling plan for the algorithm and then adjust the scheduling to allow cells to be mapped onto the finite number of cells that are required.

- The second approach is to transform the SFG representation of the circuit to allow the required circuit implementation to be achieved. This involves mapping of the SFG - a process which is best done visually.

This first approach has the advantage that the problem can be automated using software. It has the disadvantage in that the user finds it difficult to visualize the final solution and thereby direct the synthesis flow. The second process does allow visualization and, in this chapter, two different mappings are presented. On the negative side, it is difficult to envisage how this approach would be automated and existing tools do not exist for this purpose. This RLS filter example is very appropriate as it

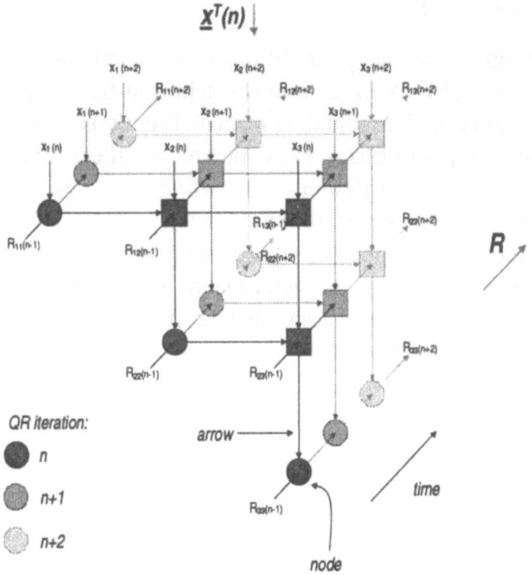

Figure 5.8 Three dimensional DDG

involves the development of a linear array solution from an initial triangular array. In this chapter, the latter approach is adopted for illustration purposes.

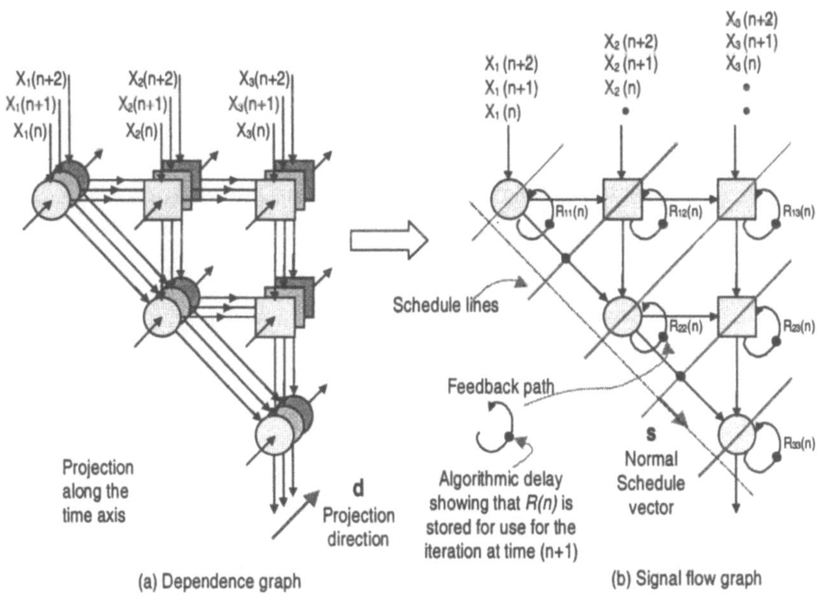

Figure 5.9 Projection of the QR processor

5.5.1 QR processor

One process for mapping the triangular QR processor onto a linear array of processors is illustrated in figure 5.10. A linear array is achieved by mapping the cells in the triangular array in either the horizontal or vertical directions. However, this results in an implementation with poor utilization as shown. An alternative approach is to use the mapping proposed by Radar [Rader, 1996] which is shown in figure 5.11. In this approach, the triangular array is partitioned as shown in figure 5.11(a). The larger part of the array is inverted and mapped as shown in figures 5.11(b), (c) and (d). Whilst the circuit results in a linear array, the requirement of the mapping is that the cells in the new architecture must be able to compute both functions. This has implications at the operation/instruction level part of the process.

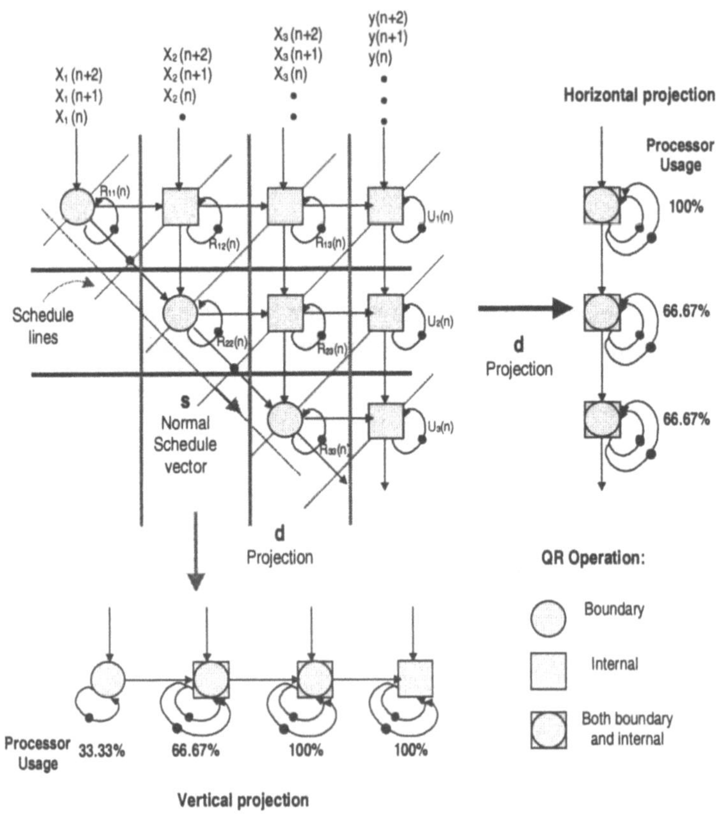

Figure 5.10 Basic projections of the QR processor array

An alternative approach by Walke [Walke, 1997] is illustrated in figures 5.12 - 5.15 and shows a different partitioning of the SFG. This results in a linear array with 100% utilization. However, the main difference between this approach and that proposed by Radar [Rader, 1996] is that it has achieved 100% utilization whilst still preserving the separate cell functionality. This is achieved by cutting the array as shown in fig-

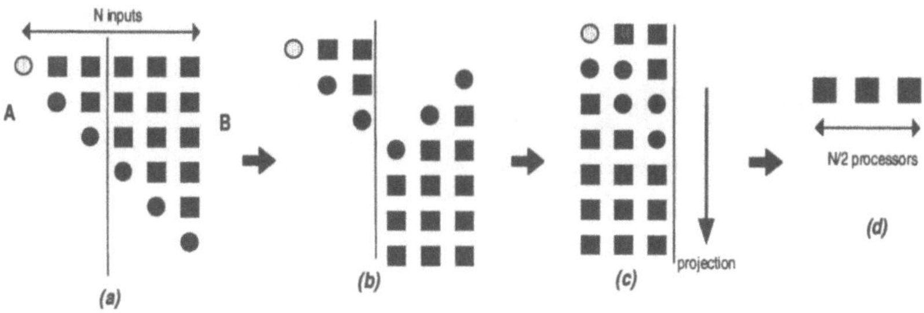

Figure 5.11 Radar's projection for the QR processor

ure 5.12 and then mirroring the lower (B) triangular matrix as shown. The B matrix is then moved as shown in figure 5.13. The parallelogram is split as shown in figure 5.14 and the cells are interleaved by folding over the cells on the right hand side of the parallelogram(figure 5.15). With this representation, a simple collapsing of operations as discussed earlier, is then applied to achieve a linear array with 100% utilization. It should be noted that the solution requires that several QR computations are overlapped but this does not present a major problem. Lightbody [Lightbody, 2000] has extended this approach and shows that rectangular and sparse array architectures are also possible. This provides highly efficient solutions that give 100% utilization for most of the input and throughput rate requirements.

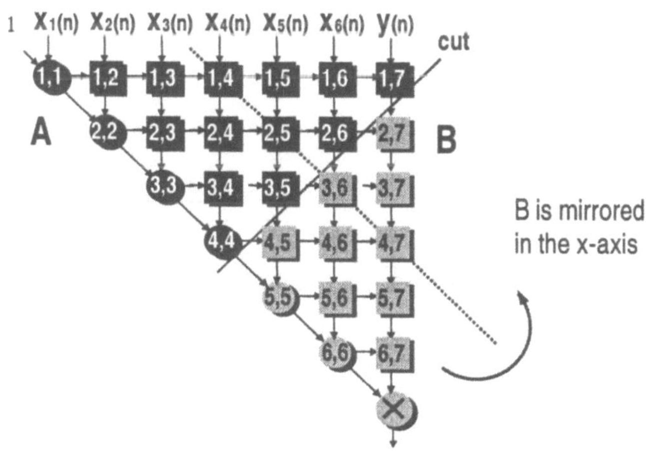

Figure 5.12 Dividing the QR array into two smaller triangular parts

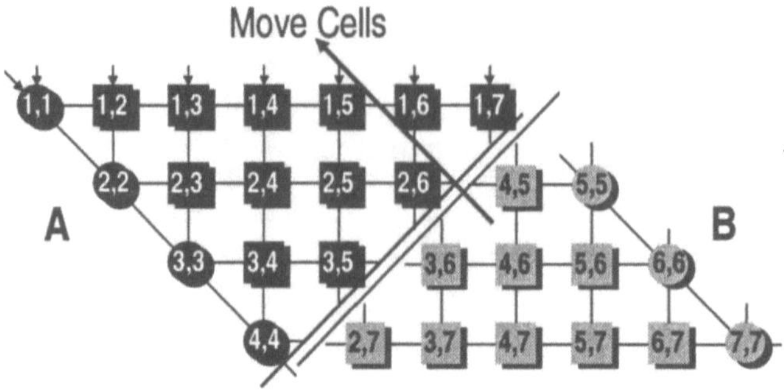

Figure 5.13 Parallelogram shape resulting after first stage

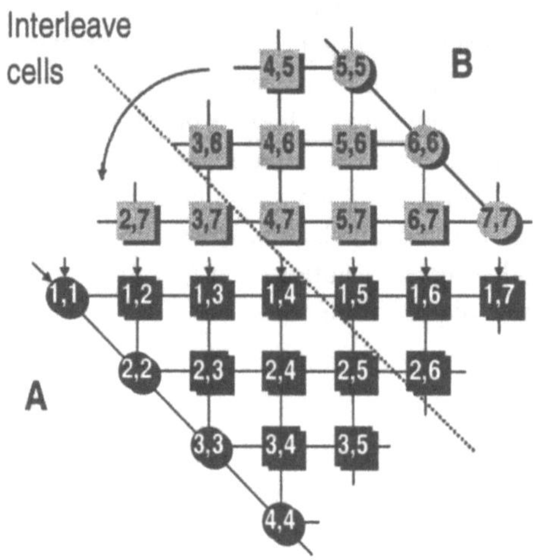

Figure 5.14 Globally connected rectangular array

5.6 OPERATION/INSTRUCTION LEVEL ISSUES

At this stage in the design flow, the development of a candidate VLSI architecture will have been chosen. A simple approach is to instantiate a processor for each of the processing elements. There are a variety of options open to the designer ranging from the arithmetic used for the instruction level processors to the choice of an operation level processor. Issues addressed at this level include:

Choice of arithmetic style: This includes the use of alternative number systems such as redundant and residue arithmetic representations. Residue arithmetic works well in applications where word growth is limited, so it has not had a major im-

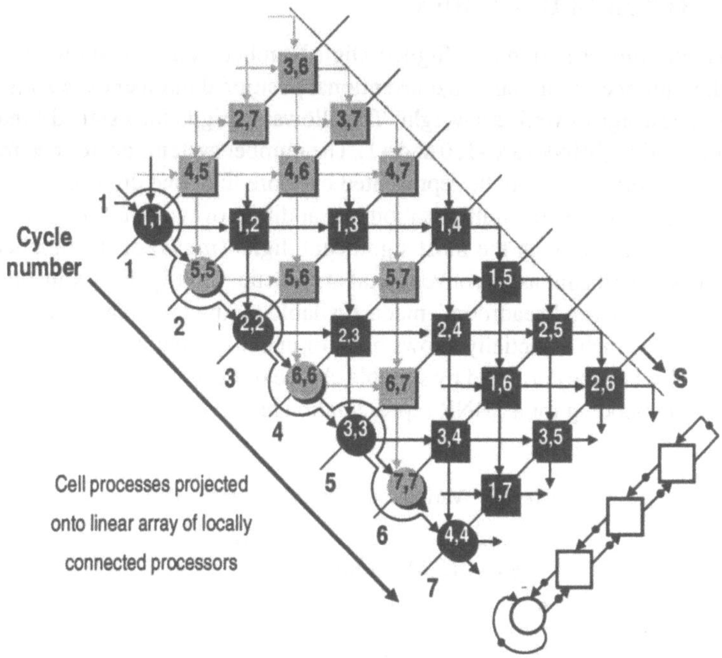

Figure 5.15 Interleaved processor array

pact in DSP. Redundant arithmetic has particular advantages from a low power perspective in that it breaks the carry propagation chains in addition and therefore multiplication. This considerably reduces the circuit switching and should be investigated from a low power perspective.

Detailed architecture of the instruction level processor: A hierarchical approach based on simple, processors can be used. Alternatively, it is possible to develop the circuit architecture using, for example, an approach based on the Co-ordinate Rotation Digital Computer (CORDIC) transformation. This approach would have to be employed in the circuit architecture by Radar [Rader, 1996] as that implementation required that the new cell performed the functionality of both cells. The choice of bit level circuit architecture can also have a major impact on power performance.

An important part of the synthesis process is to retime the circuit architecture. In many cases, the delays and datapath representations of the processors will corrupt the original functionality and delays must be added in a minimal manner to correct this. This process is covered in this section.

5.6.1 Redundant Arithmetic

In redundant number systems or Signed Digit Number Representations (SDNRs), a signed-digit number is similar to a conventional positional number except that each of its digits have a sign as well as a weight. The allowable digits for a signed-binary number representation (SBNR) are -1, 0 and +1. The number systems are termed redundant because any given value may be represented by more than one digit pattern. It is this redundancy that allows the multiplication and addition of two multi-digit numbers to be carried out starting with the most significant digits ($msbs$) of the numbers rather than the least significant as in conventional arithmetic. Briefly, this is made possible by exploiting the extra "headroom" made available by the redundancy in the number representation, which essentially allows over- or under-estimations in the more significant result digits to be corrected by suitable choice of the digits of lower significance. Two types of encoding for a SBNR representation are given in table 5.2.

Table 5.2 Various encoding styles for SBNR

Digit d	*(S,M)coding* $d_s\ d_m$	*(+,-)coding* $d_+\ d_-$
0	0 0	0 0
1	0 1	1 0
-1	1 1	0 1
0 or x	1 0	1 1

This SBNR representation allows the design of adders where carry propagation is eliminated. The attraction of the (+,-) encoding is that it is achieved using conventional bit adder cells. This is done by allowing various interpretations of the adder as shown in figure 5.16.

The circuit to add two (+,-) coded signed binary operands, a=(a^+,a^-) and b=(b^+,b^-) is shown in figure 5.17. The bit slice of the adder comprises two full adders operating sequentially. The first stage full adder accepts an SBNR digit b and the bit a^+ and generates a negatively weighted interim sum bit and a positively weighted interim transfer bit. The second stage full adder sums three bits namely the input bit a^-, the interim sum from the first adder and the interim transfer from the adjacent first stage full adder. The adder then generates a positively weighted sum and a negatively weighted transfer bit. No further addition is required. The main advantage of this approach is that there is no propagation of carries in the circuit. In this way, totally parallel addition is achieved and msb operation is possible if required.

The power consumption of this circuit will obviously be greater than the power consumption of a single adder as twice the number of adders are needed. Therefore, the power gained in removing the carry propagation chain comes at the expense of additional area and power. However, there are many cases in DSP where one of the operands will be constant such as in filtering and transforms e.g. DCT. In this case, it is possible to use a modified adder with one binary input and one SBNR input as shown

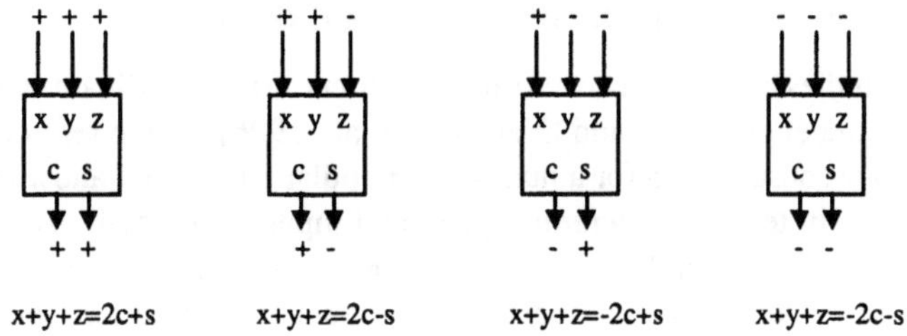

Figure 5.16 Interpretations of full adders

in figure 5.18. The advantage of this circuit is that it has the same complexity as a conventional binary adder but does not have any carry propagation. More importantly, the power consumption is reduced when compared to the equivalent carry-ripple adder. It should be pointed out that redundancy overflow must be addressed. This is where additional bits typically 2 bits for SBNR systems are generated at the msb positions due to the arithmetic used.

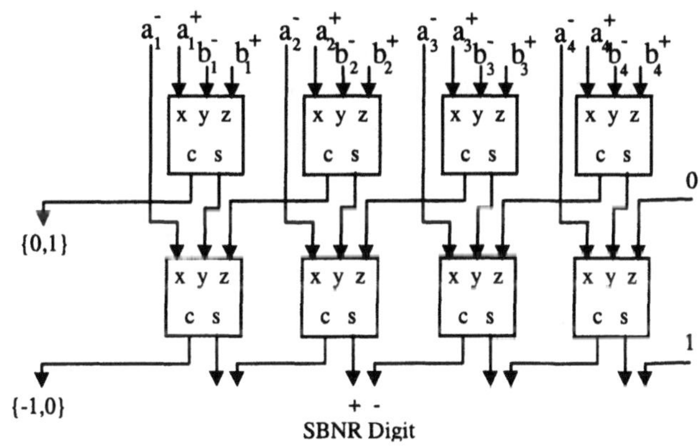

Figure 5.17 SBNR/SBNR adder

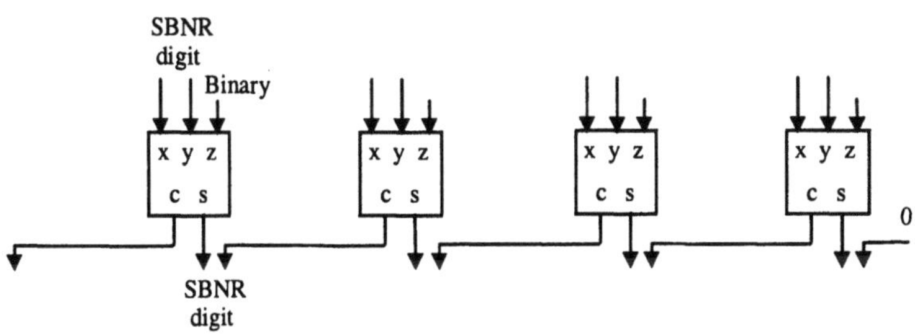

Figure 5.18 SBNR/Binary adder

5.6.2 Power Estimation of Processor Block

A number of studies have been carried out into adders structures [Shams and Bayoumi, 1997], [Nagendra et al., 1994] and [Nagendra et al., 1996]. These references provide details of power consumption for a number of popular adders and allow the adders to be characterised in terms of operating speed and input wordlength. A similar study has been carried out into multiplier structures [Keane et al., 1998] and [Keane et al., 1999] which gives details of a number of commonly used multipliers generated using a 0.35μ standard cell CMOS technology. The multiplier structures examined were: a carry-save array multiplier [Deverell, 1975]; a Booth-encoded [Booth, 1951] version of the carry-save multiplier and; a Booth-encoded Wallace tree multiplier [Fadavi-Ardekani, 1993]. Two's complement versions of the structures (designated 2C) were also examined.

The carry-save architecture (figure 5.19) comprises an array of gated full adder cells where the carry and save bits are passed vertically down through the array and are finally summed using a fast adder. Other array architectures [Deverell, 1975], [Mc-Canny and McWhirter, 1982] are available, but their critical paths tend to be longer. The Wallace tree multiplier [Wallace, 1964], shown in figure 5.20, uses carry-save adders and a fast adder to generate the final result and has increased irregular routing compared to the carry-save structure. Two additional structures, created by applying the Booth's algorithm to the carry-save and Wallace tree structures, were also considered. All of the structures were chosen as they are commonly used to implement DSP functions.

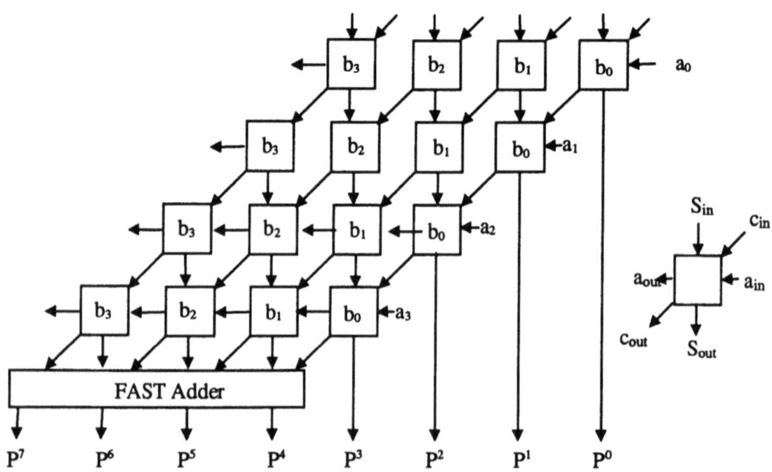

Figure 5.19 Analysis of multipliers

The multiplier structures presented all have different capabilities e.g. the Wallace-Tree structures can operate at much higher frequencies than the array structures. This factor was taken into consideration when comparing the hardware units by optimizing

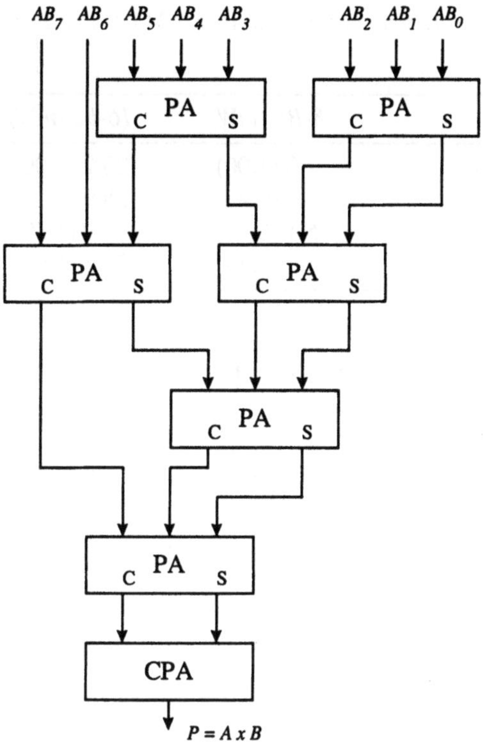

Figure 5.20 Analysis of multipliers

the designs for the required operating speeds to ensure a consistent power consumption comparison.

The power consumption of the multipliers varies with wordlength as shown in table 5.3. It indicates that the carry-save multiplier has lower power consumption than the Wallace tree. Preserving hierarchy at the circuit level (as indicated by the *regular* label) allows an even more greater saving than version where layout was not preserved (labelled *flat*). The savings in terms of both area and power grows as the wordlength increases. The reasons for this can be seen in terms of the area figures given in table 5.4 and layouts given in figures 5.21- 5.24.

Low power processor configurations. The use of transformations such as parallel processors and pipelining to reduce power is well understood [Chandrakasan and Brodersen, 1996]. Typically the transform is applied to achieve a speed-up beyond that required. The extra performance is then traded-off by decreasing the voltage. As power is dependent on the square of the voltage, a substantial power saving is achieved. An example of the power savings achieved using these architectural techniques has been presented [Chandrakasan et al., 1992]. In this research, comparative results for a very simple datapath consisting of an 8-bit adder and comparator have

Table 5.3 Multiplier power consumption at 8, 16 and 24 bit wordlengths processing random data at 20MHz

Multiplier Structure	*8-Bit (mW)*	*16-bit (mW)*	*24-bit (mW)*
Carry-save (Regular)	7.0 (1.00)	23.2 (1.00)	56.9 (1.00)
Carry-save (Flat)	3.2 (0.46)	25.9 (1.12)	68.0 (1.19)
Booth Encoded Carry-save	5.1 (0.73)	27.9 (1.20)	88.6 (1.56)
Booth Encoded Wallace tree	5.5 (0.79)	37.1 (1.60)	93.6 (1.65)

Table 5.4 Silicon area of multipliers at 8, 16 and 24 bit wordlengths optimized for 20MHz operation

Multiplier Structure	*8-Bit (mm2)*	*16-bit (mm2)*	*24-bit (mm2)*
Carry-save (Regular)	0.05	0.20	0.45
Carry-save (Flat)	0.02	0.12	0.29
Booth Encoded Carry-save	0.30	0.12	0.33
Booth Encoded Wallace tree	0.03	0.14	0.35

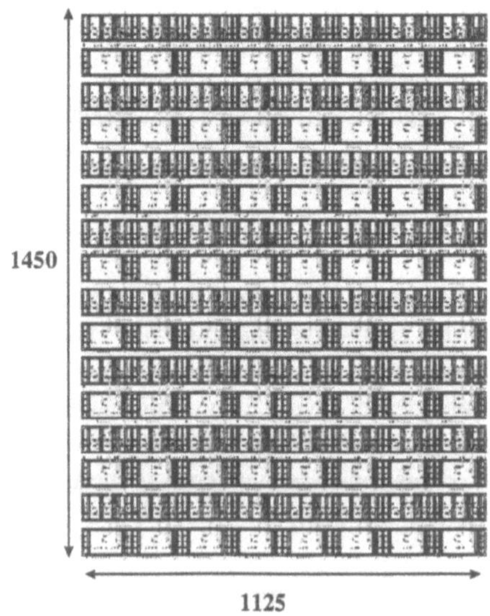

Figure 5.21 8 bit Carry-save multiplier layout(dimensions in lambda)

been presented. Using two datapaths, a 64% saving was reported over a single datapath. Using a two stage pipeline, a 61% saving was achieved. Combining both techniques resulted in an 80% overall power reduction.

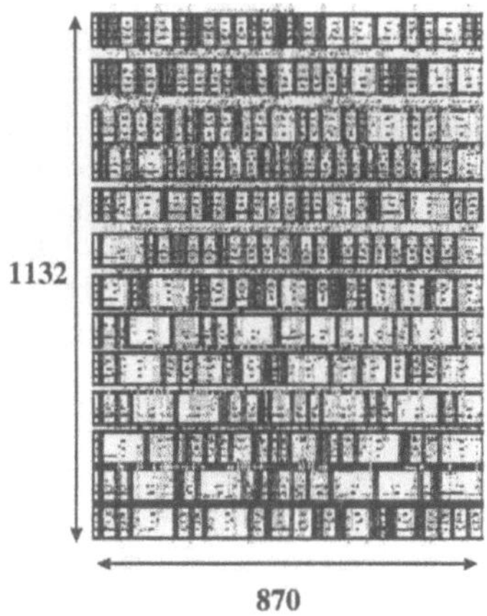

Figure 5.22 8 bit Wallace tree multiplier layout(dimensions in lambda)

While these techniques have been used with voltage scaling, they have been applied here in the standard cell flow where voltage cannot be changed. The aim of the work is to determine the power efficiency of using a very irregular but very fast multiplier, e.g. a Wallace tree multiplier, and compare it to a pipelined core such as a pipelined carry-save array. Once again circuit layouts were generated and results (table 5.5) obtained from PowerMill^{TM} simulations using one random data input of 1000 test vectors and one fixed input.

Table 5.5 Power consumption of transformed 16-bit carry-save multiplier processing random data with one fixed input at 20MHz

Multiplier Structure	*Power (mW)*	*Power-Area Product (mWmm$_2$)*
Carry-save (Flat)	9.2	1.1
Two Carry-save in parallel	7.3	3.4
Four Carry-save in parallel	12.6	15.6
Pipelined Carry-save (one pipeline)	11.6	1.7
Pipelined Carry-save (two pipeline)	14.1	2.5
Booth Encoded Wallace tree	20.3	2.8

The parallel implementation achieved a 21% power reduction over the single implementation at a considerable area cost - indicated by the power-area product. The pipelined version of the circuits faired less well but resulted in a considerably less area increase. More importantly, a pipelined carry-save structure results in a power consumption less than the Booth encoded Wallace tree structure. The Wallace tree structure consumes 43% more than the pipelined carry-save multiplier. This has implications for choosing hardware for speed-critical applications. It can be seen that

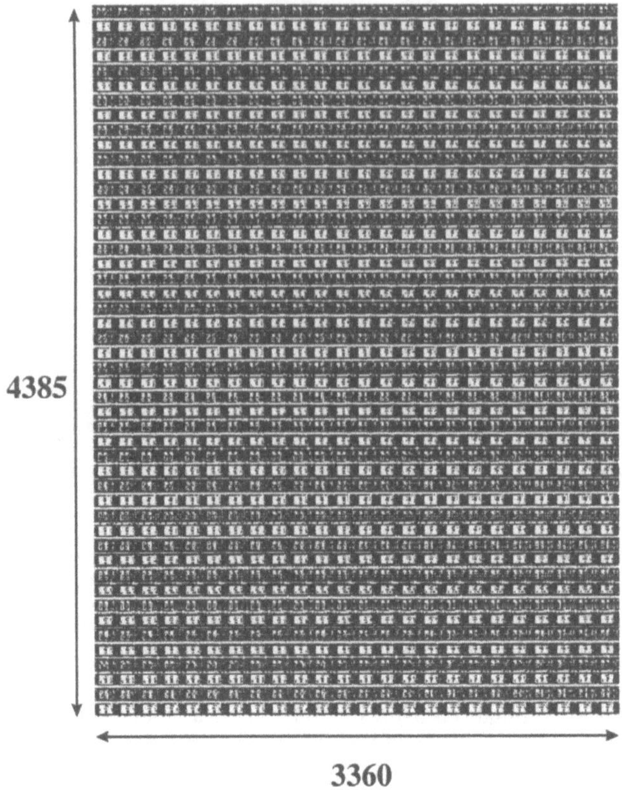

Figure 5.23 24 bit Carry-save multiplier layout(dimensions in lambda)

pipelined regular architectures will offer power savings over other structures while still providing the required speed.

5.6.3 Retiming and Rescheduling of bit level processing elements

When the choice of simple processor block is made, the individual processing blocks must be fed into the architecture and their effects evaluated. These effects are primarily timing effects such as datapath latencies and timing skew between input bits. This latter effect occurs in heavily pipelined multipliers where data may emerge least significant bit (lsb) or even msb first. A number of popular data time shapes are shown in figure 5.25. Obviously connecting blocks together with different time shapes will create timing issues and cause the circuit to malfunction. A number of approaches can be adopted to avoid this problem. The user can avoid the problem by keeping to one data time format which will avoid many of the problems but will severely restrict the options for the designer. Alternatively, the user can apply retiming techniques.

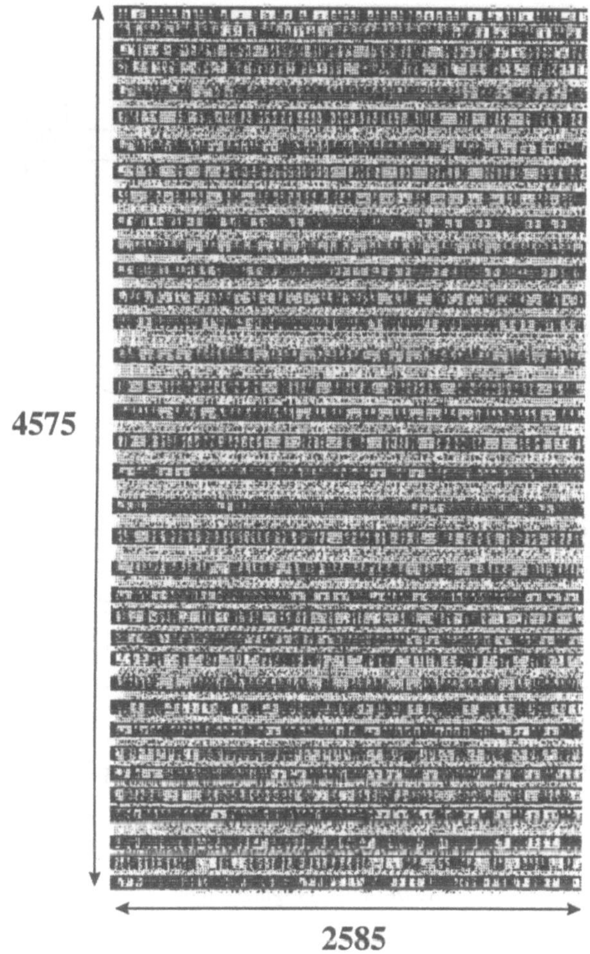

Figure 5.24 24 bit Wallace tree multiplier layout(dimensions in lambda)

The problem is highlighted in the simple structure of figure 5.26 which shows two multiply-accumulate blocks that have been connected together. The mathematical relationship to be implemented is given as:

$$y_n = a_2.y_{n-1} + a_1.y_{n-2} + x_n \quad (5.4)$$

The original SFG assumes zero delays and algorithmic delays must be explicitly stated. In the case of figure 5.26, the black dots represent these delays. Embedding the processors as shown in figure 5.27 means that the original functionality is now lost unless additional delays are included as shown. This situation is considerably complicated if there are input and output timing conflicts where data with different time shapes needs to be connected together.

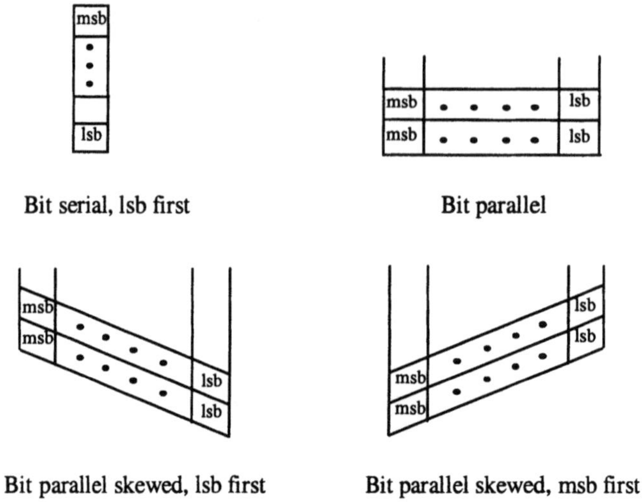

Figure 5.25 Examples of typical data-time shapes

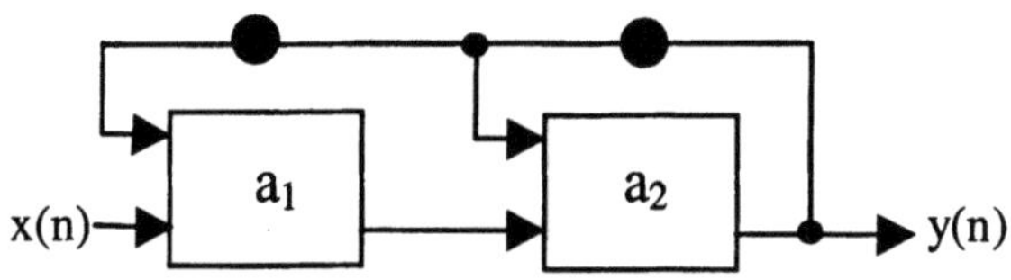

Figure 5.26 Original circuit

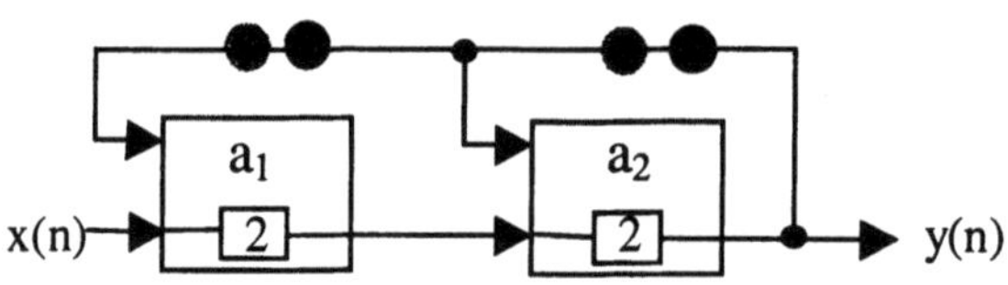

Figure 5.27 Retimed circuit

The insertion of these delays can be achieved in a number of ways. One example is to apply the cut theorem proposed by Kung [Kung, 1988]. Briefly, a cut can be considered as a minimal set of SFG edges that partitions the SFG into two parts (figure 5.28). The retiming is based on two simple rules:

time scaling: The number of delays present on the edges of a SFG may be scaled by a single positive integer without altering the overall timing of the SFG.

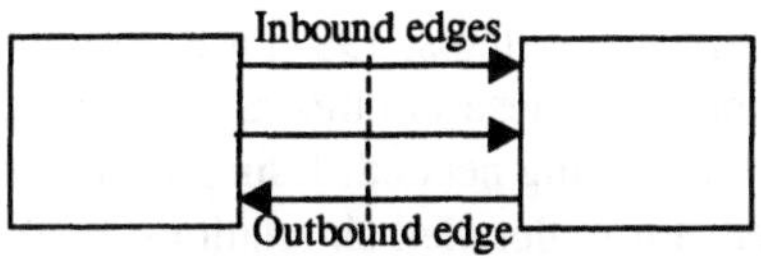

Figure 5.28 Definition of the cut theorem

Delay transfer: After the application of the cut, the group or edges that have been cut ($cut - set$) can be divided into two groups. These groups are the inbound edges and outbound edges as illustrated in figure 5.28.

The transfer rule allows a number of delays to be transferred from outbound to inbound edges without affecting the global timing. In recursive structures such as the example of figure 5.27, the worst case time scaling must be calculated and the circuit architecture delayed accordingly. This positive integer is the pipelining period and can be referred to as α_c. The optimal pipelining period is defined by equation 5.5.

$$\alpha = max(all\alpha_c) \tag{5.5}$$

If B_C is deemed to be the delays in the loop due to the pipelining of the original SFG and D_C is the total number of delay operations in the cycle C, then α_C is given by equation 5.6.

$$\alpha_c = \lceil \frac{B_c}{D_C} \rceil \tag{5.6}$$

B_C refers only to the delays attributable to the processor pipelining and not to any pure delays that will be needed for correct timing. In the example of figure 5.27 there are two loops. One loop is around the second multiplier-accumulate block and has a B_C value of 2 and a D_C value of 1. The second loop has a B_C value of 4 and a D_C value of 2. Therefore α_c is equal to 2 in both cases. Using this value, the circuit is retimed using the cut-theorem to achieve the final version. This technique has been incorporated within a synthesis tools IRIS [Trainor et al., 1997]. By effectively calculating the worst case loop and moving flipflops about efficiently, a low power solution is achieved where the flipflops are minimized. A more detailed description of the synthesis process is given in Trainor et al. [Trainor, 1995].

5.6.4 Application to RLS filter instruction level implementation

For the RLS cells given earlier in figure 5.6, the cell functionality can be broken down into simple, arithmetic processors. The instruction level process involves instantiating various processors for each of the functions to give the cell implementations of figure 5.29. For this example, a commercially available arithmetic library called

DSiPWare^{TM} has been used to develop the cell descriptions. These DSiPWare^{TM} library cores [Integrated Silicon Systems, 1997] are parameterizable VHDL descriptions that can be programmed. These circuits are parameterizable in terms of data wordlengths and levels of pipelining needed. Using these cores, detailed cell descriptions have been developed. From detailed simulations for the adaptive beamforming applications, a floating point mantissa of 12 bits and an exponent of 5 bits were deemed suitable. The target of this research was to produce a design with the ultimate performance namely a clock rate of 100MHz in a 0.35μm standard cell CMOS technology. This involves 4 levels of pipelining for the complex operators such as division and multiplication. The power consumption of these processors are shown in table 5.6.

These cores were then inserted into the cell descriptions. A latency of 7 cycles is determined for the loop in the internal cell (using the synthesis tool IRIS) which was then used to retime the circuit by adding delays as required. This is discussed in detail in Lightbody[Lightbody, 2000]. The resulting design comprises 140,000 transistors and the power consumption of the internal cell is given as 1.27W.

An alternative approach is to break down the operations into simpler processing blocks, as shown in figure 5.30. In this design, a series expansion has been used to implement the division operation and individual multipliers and adders have been used to implement the complex multiplications, thereby reducing all operations to multipliers and adders.

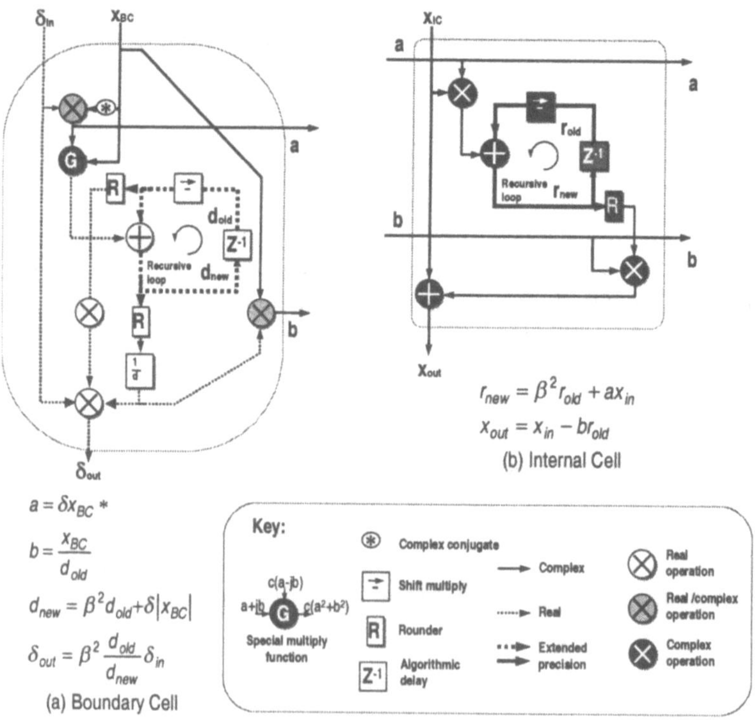

Figure 5.29 QR processor cell descriptions

This approach has two advantages. As can be seen from table 5.6, the floating point divider has a power consumption that is nearly four times the power consumption of a floating point multiplier. Therefore, if the division operations are replaced by multiplications, then this could reduce both the area and the power consumption. This then allows another level at which hardware mapping can be performed as the individual hardware units in the cell can now be shared. This would be useful in applications where lower throughput rates are required.

Table 5.6 Power Consumption analysis of DSiPWareTM cores simulated at 100MHz(All Floating point)

Component	*Wordlength (bits)* *Mantissa*	*Exponent*	*No. of Transistors*	*Power (mW)*
Adder	12	5	6556	42.3
Subtracter/Adder	12	5	6408	43.3
Multiplier	12	5	8524	64.5
Shifter Subtracter	12	5	1644	9.8
Divider	12	5	24434	227.1

5.7 1D DCT EXAMPLE

The DCT is a key DSP function that has been used extensively for image processing and some speech coding applications. The 2D version of the DCT shown is a separable function that can be split into two single DCT functions. This approach is commonly termed row-column decomposition. The 1D DCT equation is shown in equation 5.7.

$$Y(k) = \sqrt{\frac{2}{N}}\alpha(nk)\sum_{n=0}^{N-1} x(n)cos\frac{(2n+1)\pi k}{2N} : k = 0, 1, ..., N-1 \qquad (5.7)$$

From this analysis, a number of algorithmic approaches can be applied, two of which are demonstrated here.

5.7.1 Main DCT implementation

The first approach involves the transformation of equation 5.8 into a transfer function resembling a second order IIR filter representation. This is shown below and has been derived by Hunter [Hunter, 1999].

$$H_k(Z) = \sqrt{\frac{2}{N}}\alpha(k)cos(\frac{\pi k}{2N}).cos\frac{(2n+1)\pi k}{2N} : k = 0, 1, .., N-1 \qquad (5.8)$$

A VLSI circuit for this architecture is shown in figure 5.31. The main advantage of this approach is that the architecture is generic and can allow a number of solutions to

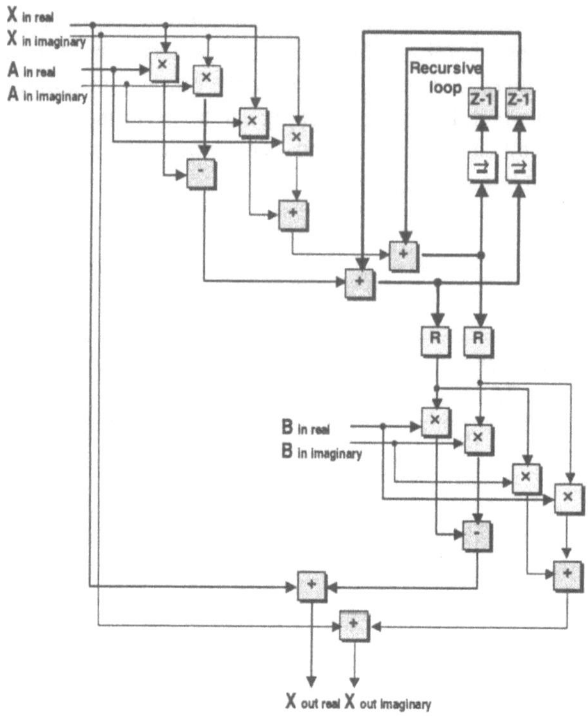

Figure 5.30 Revised internal cell description

be derived with varying wordlengths and point sizes. Different performance requirements and processing speeds are achieved by cascading the block as required. In the specific circuit of figure 5.31, the design uses a multiplier that has been pipelined 4 times and the final circuit has been retimed to reflect this. The internal wordlengths have been determined by detailed simulations for the specific JPEG application. The initial design used Wallace tree multipliers and carry lookahead adders.

The specification outlined by Hunter [Hunter, 1999] required a circuit with a 8 bit input, 11 bit output and a 12 bit internal datapath and that operates at 100 MHz. The M1 coefficient is 12 bits whilst M2 is 10 bits. This initial circuit was optimized for speed. A number of approaches were investigated. The hierarchical approach involved the derivation of a solution where the individual processing blocks multipliers and adders are synthesized and implemented individually. An approach using carry-save multipliers was investigated where the circuit layout is generated without imposing any hierarchy i.e. flat. A hierarchical DCT approach using carry-save multipliers was also investigated. These results are shown in table 5.7 and graphically in figure 5.32. Table 5.7 shows the power reduction achieved by selecting a carry-save multiplier over a Wallace tree for a DCT implementation. It should be noted that very little effort was expended in generating the layout for the hierarchical examples. This is reflected

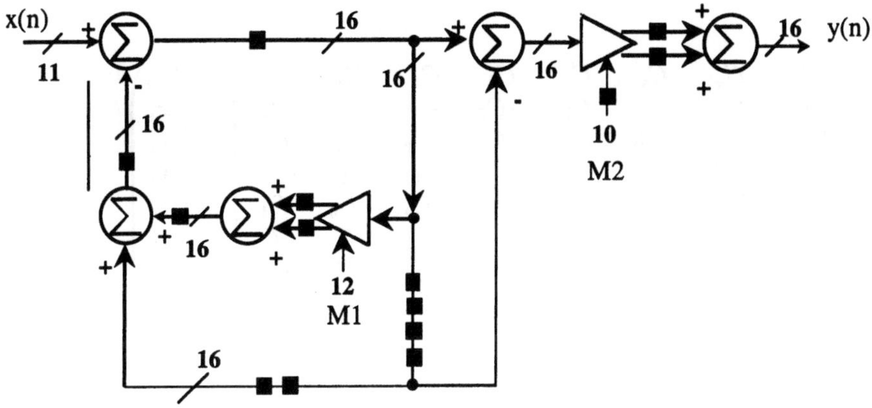

Figure 5.31 Details of the parameterizable DCT core

by the transistor density figures. Considerable area improvement and further power reductions should be possible.

5.7.2 Alternative DCT implementation

Whilst this previous approach gives a good solution across a range of parameters, it is possible to derive a power-efficient solution by optimizing the design for a particular point size. This work relies heavily on the existing body of work that has been carried out into minimizing the matrix-vector representation of the 1D DCT. The matrix vector representation for 1D DCT equation given earlier, is shown in equation 5.9. The expression of equation 5.9, represents the generation of one row of the internal DCT matrix using one row of the input matrix and will therefore be performed 8 times. This internal DCT matrix will then be transposed and the operation repeated to give the full 2D DCT operation. Note that the scaling has been ignored as it can be realized as a simple shifter in the 2D DCT circuit.

Table 5.7 Power consumption of different DCT core architectures processing image data

Architecture	*Power (Image) (mW)*
Original DCT System (Wallace tree)	64.7 (1.00)
Hierarchical DCT System (Wallace tree)	58.9 (0.91)
Flat DCT system with carry-save Multipliers	49.0 (0.76)
Hierarchical DCT with carry-save Multipliers	44.0 (0.68)

$$\begin{pmatrix} C_4 & C_4 & C_4 & C_4 & C_4 & C_4 & C_4 & C_4 \\ C_1 & C_3 & C_5 & C_7 & -C_7 & -C_5 & -C_3 & -C_1 \\ C_2 & C_6 & -C_6 & -C_2 & -C_2 & -C_6 & C_6 & C_2 \\ C_3 & -C_7 & -C_1 & -C_5 & -C_5 & C_1 & C_7 & -C_3 \\ C_4 & -C_4 & -C_4 & C_4 & C_4 & -C_4 & -C_4 & C_4 \\ C_5 & -C_1 & C_7 & C_3 & -C_3 & -C_7 & C_1 & -C_5 \\ C_6 & -C_2 & C_2 & -C_6 & -C_6 & C_2 & -C_2 & C_6 \\ C_7 & -C_5 & C_3 & -C_1 & C_1 & -C_3 & C_5 & -C_7 \end{pmatrix} \begin{pmatrix} x_0 \\ x_1 \\ x_2 \\ x_3 \\ x_4 \\ x_5 \\ x_6 \\ x_7 \end{pmatrix} \quad (5.9)$$

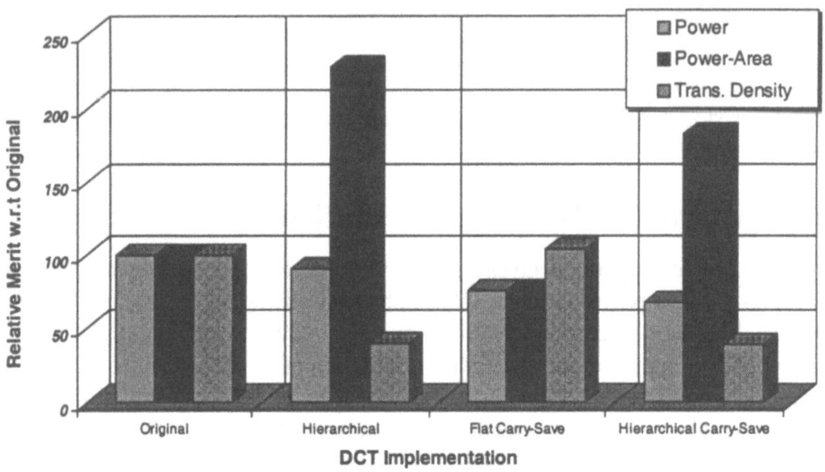

Figure 5.32 DCT Analysis

The values C_x are DCT coefficients and are given by $cos(x/16)$ where x = 1..7. As it stands, 64 multiplications (and 64 additions) which corresponds to 512 multiplies (and 512 additions) are required for the full 8X8 block. Each multiplication corresponds to one non-zero coefficient value in the coefficient matrix. However, there are only 7 different coefficient values in the coefficient matrix which allows considerable scope to reduce the number of multiplications. By exploiting the fact that the DCT coefficient matrices are fixed, the number of multiplications can be considerably reduced. This is done by manipulating the input data at the cost of addition and subtraction operations to get a reduction of multipliers. One example of this is shown in equation 5.10.

$$
\begin{pmatrix} y_0 \\ y_2 \\ y_4 \\ y_6 \\ y_1 \\ y_3 \\ y_5 \\ y_7 \end{pmatrix} = \begin{pmatrix} C_4 & C_4 & C_4 & C_4 & 0 & 0 & 0 & 0 \\ C_2 & C_6 & -C_6 & -C_2 & 0 & 0 & 0 & 0 \\ C_4 & -C_4 & -C_4 & C_4 & 0 & 0 & 0 & 0 \\ C_6 & -C_2 & C_2 & -C_6 & 0 & 0 & 0 & 0 \\ 0 & 0 & 0 & 0 & C_1 & C_3 & C_5 & C_7 \\ 0 & 0 & 0 & 0 & C_3 & -C_7 & -C_1 & -C_5 \\ 0 & 0 & 0 & 0 & C_5 & -C_1 & C_7 & C_3 \\ 0 & 0 & 0 & 0 & C_7 & -C_5 & C_3 & -C_1 \end{pmatrix} = \begin{pmatrix} x_0 + x_7 \\ x_1 + x_6 \\ x_2 + x_5 \\ x_3 + x_4 \\ x_0 - x_7 \\ x_1 - x_6 \\ x_2 - x_5 \\ x_3 - x_4 \end{pmatrix} \quad (5.10)
$$

Here the multiplications have been reduced from 64 to 32. Further reduction is possible and there has been a considerable body of work into reduced DCT coefficient matrix solutions. Table 5.8 presents details of the number of multiplication and additions for the key research papers. Whilst this demonstrates that further reduction in number of multiplications is possible, there is a trade-off between the further reduction in the number of multipliers and the regularity of the circuit layout. Therefore, it is preferable to use the representation of equation 5.10 as a starting point as the resulting architecture is highly regular (figure 5.33). Another reason for considering the expression of equation 5.10 is that a bit serial implementation is being considered. This allows each new computation of the matrix to be realized as LUTs rather than individual multipliers and adders and these can be synthesized as gates. In addition, the transform circuitry needed to re-organize the DCT values for the second 1D DCT computation can be realized by a simple MUX circuit rather than a solution based on memory. This is shown in figure 5.34. The resulting design can be pipelined as required to meet the performance requirements. This latter approach will give a low power consumption for specific designs but cannot be parameterized to the levels demonstrated in the first circuit.

5.8 CONCLUSIONS

The chapter has given some insights and power consumption figures for the design of custom processors for regular digital signal processing modules. From a power perspective, it is important to develop a coherent strategy for the synthesis of highly regular structures. This involves the development of algorithmic synthesis methods to efficiently map the initial mathematical equations into a dependence graph (DG) representation where regularity and the possibility of achieving a highly efficient signal flow-graph (SFG) representation is possible further down the design trajectory. At the SFG level (which is part of the data/array-level abstraction), it is important

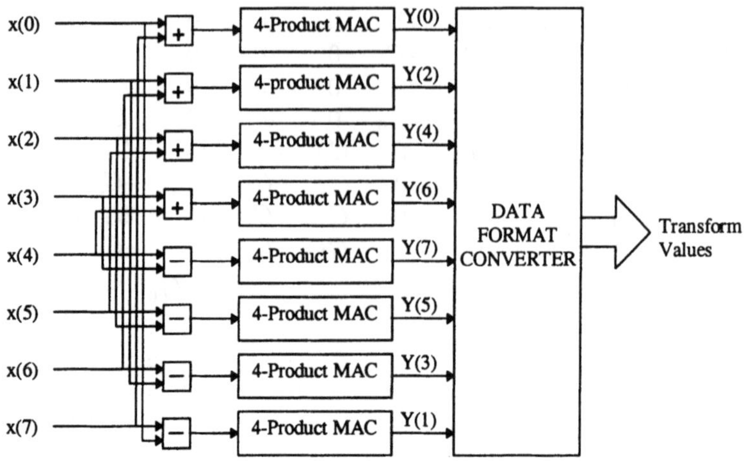

Figure 5.33 Matrix vector DCT circuit

Table 5.8 Comparison of fast algorithms

Fast Algorithm	*No. of multiplications*	*No. of Additions*
Chen et al.[Chen et al., 1988].	16	26
Hou[Hou, 1987].	14	30
Cvetkovic and Popovic. [Cvetkovic and Popovic, 1992].	12	29
Loeffler et al. [Loeffler et al., 1989].	11	22

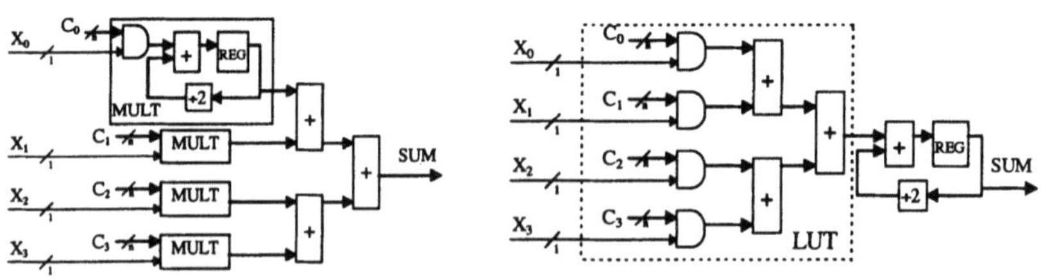

Figure 5.34 Detailed implementation of the 4 multiplier MAC

to investigate efficient array mappings that will result in 100% utilization. This can have implications for the implementation of the actual processors as highlighted by the RLS example. It was shown that the approach used by Radar [Rader, 1996] resulted

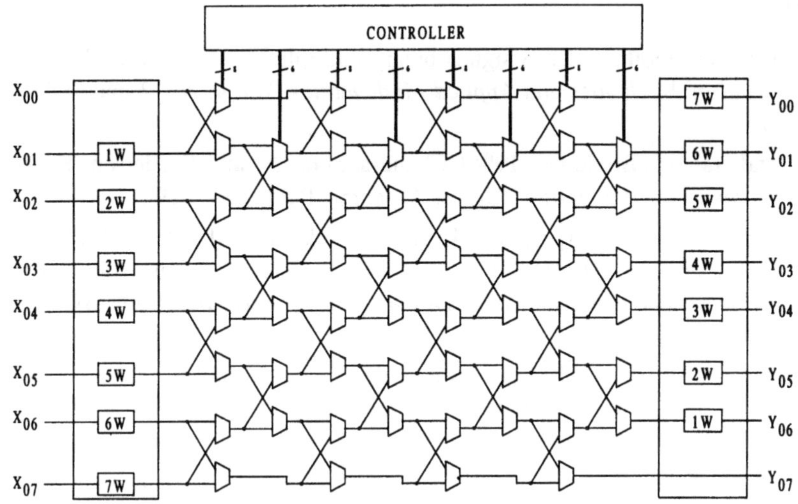

Figure 5.35 Circuit architecture for (8x8) matrix transpose design

in 100% utilization but requires the development of an processor than can perform both operations. In this case, a CORDIC implementation can be used to achieve this requirement. Initial studies suggest that this may result in a larger implementation for this application.

At the operation/instruction abstraction level, a number of alternative approaches are open to the user including the use of redundant arithmetic representations. Arithmetic processors implemented using this approach have limited carry propagation and can therefore result in lower power consumption. However, the requirements of overflow management and the need for conversion circuitry must also be considered. A processor using SBNR representation throughout may result in lower power but if there is a requirement to convert to and from two's complement representations too often, then this approach could be prohibitive. In addition, the chapter presents some results on the variation in power consumption for the custom processor architecture at the operation level. In this case, a multiplier case study is given.

The results, particularly when applied to the DCT circuit in the last section, illustrates that substantial power savings can be made with little effort. Overall, use of these techniques in data dominated applications such as image processing suggest that substantial savings are to be achieved. The availability of standard cell libraries and the development of IP cores have considerably accelerated development times. The development of high level IP cores which result in a lower power circuit implementation is a research field that is being actively pursued.

References

[Booth, 1951] Booth, A. D. A signed binary multiplication technique. In *Quarterly Journal of Mechanics and Applied Mathematics, Vol. 4, No. 2*, pages 236–240, 1951.

[Chandrakasan and Brodersen, 1996] Chandrakasan, A. and Brodersen, R., editors *Low Power Digital Design*. Kluwer Academic Publishers, 1996.

[Chandrakasan et al., 1995] Chandrakasan, A., Potkonjak, M., Mehra, R., Rabaey, J., and Brodersen, R. (1995). Optimizing Power Using Transformations. In *IEEE Transactions on Computer Aided Design of Integrated Circuits and Systems, Vol. 14, No. 1*, pages 12–31.

[Chandrakasan et al., 1992] Chandrakasan, A., Sheng, S., and Brodersen, R. (1992). Low Power CMOS Digital Design. In *IEEE Journal of Solid-State Circuits, Vol. 27, No. 4*, pages 473–484.

[Chen et al., 1988] Chen, T., Sun, M., and Gottlieb, A. (1988). VLSI Implementation of a 16x16 DCT. In *Proceedings ICASSP 1988, Vol. 4*, pages 1973–1976.

[Cvetkovic and Popovic, 1992] Cvetkovic, Z. and Popovic, M. V. (1992). New Fast Recursive Algorithms for the Computation of the Discrete Cosine and Sine Transforms. In *IEEE Trans. On Signal Processing, Vol 40, No 8.*, pages 2083–2086.

[Deverell, 1975] Deverell, J. (1975). Pipelined Iterative Arithmetic arrays. In *IEEE Trans. On Computers, C-24*, pages 317–322.

[Fadavi-Ardekani, 1993] Fadavi-Ardekani, J. (1993). M X N Booth Encoded Multiplier Generator using Optimized Wallace Trees. In *IEEE Transactions on VLSI Systems, Vol. 1, No. 1*, pages 120–125.

[Foty and Nowak, 1994] Foty, D. P. and Nowak, E. J. (1994). MOSFET Technology for Low-Voltage / Low-Power Applications. In *IEEE Micro*, pages 68–77.

[Givens, 1958] Givens, W. (1958). Computation of plane unitary rotations transforming a general matrix to triangular form. In *J. Soc. Ind. Appl. Math, vol. 6*, pages 26–50.

[Guorong and Yungong, 1990] Guorong, H. and Yungong, S. (1990). CORDIC Implementation of the RLS Systolic Array. In *Proceedings of ICSP*, pages 1133–1136.

[Haykin, 1986] Haykin, S., editor (1986). *Adaptive Filter Theory*. Prentice Hall.

[Hou, 1987] Hou, H. S. (1987). A Fast Recursive Algorithm for computing the Discrete Cosine Transform. In *IEEE Trans. On Acoustics, Speech and signal Processing, Vol ASSP-35, No. 10*, pages 1455–1461.

[Hunter, 1999] Hunter, J., (1999). *Rapid Design of Discrete Cosine Transform Cores for Multimedia VLSI Systems*. PhD thesis, The Queen's University of Belfast.

[Integrated Silicon Systems, 1997] Integrated Silicon Systems, (1997). *DSiPWare Fixed Point Library Databook*. Integrated Silicon Systems.

[Keane et al., 1998] Keane, G., Spanier, J. R., and Woods, R. (1998). The Impact of Data Characteristics and Hardware Topology on Hardware Selection for Low

Power DSP. In *Int'l Symp. on Low Power Electronics and Design (ISLPED'98)*, pages 94–96.

[Keane et al., 1999] Keane, G., Spanier, J. R., and Woods, R. (1999). Low-Power design of Signal Processing systems using Characterization of silicon IP cores. In *33rd Asilomar Conference on Signals, Systems and Computers, IEEE Computer Society*, pages 767–771.

[Kung, 1988] Kung, S. Y., editor (1988). *VLSI Array Processors.* Prentice-Hall.

[Kuroda et al., 1998] Kuroda, T., Suzuki, K., Mita, S., Fujita, T., Yamane, F., Sano, F., Chiba, A., Watanabe, Y., Matsuda, K., Maeda, T., Sakurai, T., and Furuyama, T. (1998). Variable Supply-Voltage Scheme for Low-Power High-Speed CMOS Digital Design. In *IEEE Journal of Solid-State Circuits, Vol. 33, No. 3*, pages 454–462.

[Lightbody, 2000] Lightbody, G., (2000). *High Performance VLSI Architectures for Recursive Least Squares Filtering*. PhD thesis, The Queen's University of Belfast.

[Loeffler et al., 1989] Loeffler, C., Ligtenberg, A., and Moschytz, G. (1989). Practical Fast 1D DCT Algorithms with 11 Multiplications. In *Proceedings ICASSP 1989, Vol. 2*, pages 988–991.

[McCanny and McWhirter, 1982] McCanny, J. V. and McWhirter, J. G. (1982). Completely Iterative Pipelined Multiplier array suitable for VLSI. In *IEE Proc., Vol 129, Pt G., No. 2*, pages 40–45.

[McWhirter, 1983] McWhirter, J. G. (1983). Recursive least squares minimisation using systolic array. In *Proc. SPIE (Real-Time Signal Processing IV), vol. 431.*, pages 105–112.

[Mehra et al., 1996] Mehra, R., Guerra, L. M., and Rabaey, J. M. (1996). Low-Power Architectural Synthesis and the Impact of Exploiting Regilarity. In *Journal of VLSI Signal Processing Systems, Vol. 13*, pages 239–258.

[Nagendra et al., 1996] Nagendra, C., Irwin, M. J., and Owens, R. M. (1996). Area-Time-Power Tradeoffs in Parallel Adders. In *IEEE Transactions on Circuits and Systems, Vol. 43, No. 10*, pages 689–702.

[Nagendra et al., 1994] Nagendra, C., Owens, R. M., and Irwin, M. J. (1994). Power-Delay Characteristics of CMOS Adders. In *IEEE Transactions on VLSI Systems, Vol. 2, No. 3*, pages 377–381.

[Rader, 1996] Rader, C. M. (1996). VLSI systolic arrays for adaptive nulling. In *IEEE Signal Processing Magazine, Vol. 13, No. 4*, pages 29–49.

[Shams and Bayoumi, 1997] Shams, A. M. and Bayoumi, M. (1997). A Structured approach for Designing Low Power Adders. In *31st Asilomar Conference on Signals, Systems and Computers, IEEE Computer Society, Vol. 1.*, pages 757–761.

[Stork, 1995] Stork, J. (1995). Technology Leverage for Ultra-Low Power Information Systems. In *Proceedings of the IEEE, Vol. 83, No. 4*, pages 607–618.

[Trainor, 1995] Trainor, D., (1995). *An Architectural Syntheiss tool for VLSI Signal Processing chips.* PhD thesis, The Queen's University of Belfast.

[Trainor et al., 1997] Trainor, D., Woods, R. F., and McCanny, J. V. (1997). Architectural Synthesis of Digital Signal Processing Algorithms using "IRIS". In *Journal of VLSI Signal Processing Systems, Vol. 16, No. 1,*, pages 41–56.

[Vanhoof et al., 1993] Vanhoof, J., Rompaey, K. V., Bolsens, I., Goossens, G., and Man, H. D., editors (1993). *High-Level Synthesis for Real-Time Digital Signal Processing*. Kluwer Academic Publishers.

[Walke, 1997] Walke, R. L., (1997). *High Sample Rate Givens Rotations for Recursive Least Squares*. PhD thesis, University of Warwick.

[Wallace, 1964] Wallace, C. S. (1964). A Suggestion for a Fast Multiplier. In *IEEE Transactions on Computers, Vol. 13, No. 1*, pages 14–17.

[Zimmermann and Fichtner, 1997] Zimmermann, R. and Fichtner, W. (1997). Low-Power Logic Styles: CMOS Versus Pass-Transistor Logic. In *IEEE Journal of Solid-State Circuits, Vol. 32, No. 7*, pages 1079–1090.

6 RUN-TIME POWER MANAGEMENT FOR LOW AND MEDIUM BIT-RATE DIGITAL RECEIVERS

Nikos D. Zervas, Spyros Theoharis, Minas Perakis and Costas E. Goutis

Dep. of Electrical and Computer Engineering, University of Patras, Greece

Dimitrios Soudris

Dep. of Electrical and Computer Engineering, Democritus University of Thrace, Greece

F. Catthoor (ed.),
Unified Low-power Design Flow for Data-dominated Multi-media and Telecom Applications, 135-157.

This chapter focuses on run-time power management, which is one of the most efficient low-power strategies, applicable at various levels of abstraction of the unified meta flow. Two run-time power management techniques applicable for the design of low and medium bit-rate digital modems are introduced. Specifically, an event-driven power management methodology and a dynamic power management technique based on frequency scaling, are presented for custom processors. They are both situated at the operation/instruction concurrency level in the unified design flow of chapter 2. The main test-vehicle is a real-life design, namely a DECT receiver developed for the purposes of the LPGD project in the European Low Power Initiative. The experimental results prove that the application of the proposed techniques introduces significant power savings.

6.1 INTRODUCTION

Nowadays, sophisticated hand-sets, with wireless communication capabilities, have entered the world market. In such applications low power consumption is of great importance to allow not only extended battery life, but also to reduce the packaging and cooling related cost. One of the most efficient low-power techniques, applicable at all levels of abstraction of the design flow, is the dynamic power management, applied at run-time based on the actual stimuli and context.

The discipline behind all related work is the selective shutdown of a resource, when it performs useless operations. Run-time power management techniques can be grouped into two categories with respect to the abstraction level that they apply. The first category includes techniques applicable at the task-level, while the second one applies at the circuit-level. More specifically, any task-level, run-time power management strategy is a task that is executed concurrently with the rest of the system, exchanges data with it, needs to store data and have certain performance constraints. Consequently, such a task must be refined with the rest of the system through an instantiated flow derived from the unified meta flow. The circuit-level power management techniques are embedded in the first steps of the procedure of mapping the processor-level system architecture to a certain (reconfigurable or application specific) technology.

In [5], task-level conventional power management approaches, such as those already in use in portable devices, are reviewed. The tasks targeted by these techniques correspond to system components such as display servers, communication interfaces etc. Each task is thought as a finite state system that can be in two states: *Active* or *idle*. In [26, 12], event-driven methods targeting the prediction of the idle states of tasks are presented. These techniques exploit the past history of active and idle intervals, in order to predict the length of an idle interval as soon as a system enter the idle state and from this to decide whether or not it is advantageous to prevent a task from executing. In [6, 23] techniques for the task-level, run-time power management based on stochastic modeling are proposed.

Efficient circuit-level techniques have been proposed in [1, 2, 3, 4, 27]. Specifically, *pre-computation* [1] relies on the idea of duplicating part of the logic with the purpose of pre-computing the output values one clock cycle earlier than required. Then if this is achieved the original logic is turned off in the next clock cycle, thus eliminating activity in the internal nodes. In contrast, *guarded evaluation* [27] does not require to synthesize additional logic to implement the shut-down mechanism; rather it exploits existing signals in the given circuit. The approach is based on placing transparent latches with an enable signal at the input of each block of the circuit that needs to be power managed. When the block must execute some useful computation, the enable signal makes the latches transparent. Finally, in [2, 3, 4] a method for the automatic synthesis of gated clocks in sequential circuits that implements reactive systems was presented. Reactive systems wait for a certain event to occur before changing state. During the wait periods the system's outputs do not change, and if the system is clocked, much power can be wasted. This method recognizes these idle states and inserts the appropriate logic that ceases the clock. An extensive review of processor- and circuit-level power management techniques can be found in [16].

In this chapter, the power management discipline is applied to custom processor implementations of low and medium bit-rate, wireless and wire-line digital receivers. The exploitation of specific characteristics of receivers belonging in this wide category, has lead to two techniques that can apply in addition to existing techniques and can lead to further power savings. Specifically, a formalized methodology for the operation/instruction-level event-driven power management and a dynamic power management technique based on frequency scaling are presented. In particular, the methodology aims at the early exploration of the trade-off according to which on the one hand power consumption decreases because a resource is shutdown but on the other hand power consumption and area increases due to the presence of additional logic. The dynamic frequency scaling is a processor-level power management technique, which uses frequency scaling to "force" the receiver to operate only on the useful data. This technique requires minor architectural changes with respect to conventional architecture styles, while negligible increases area and critical path.

6.2 BASIC BACKGROUND

The scope of this section is to provide the reader with the basic information, necessary to better understand the rest of this chapter. In the first subsection, basic telecommunication terms are defined and the synchronization scheme most commonly met in low and medium bit-rate receivers is described. The second subsection provide a brief description of the main test vehicle used throughout the rest of the chapter, which is a DECT receiver.

6.2.1 Synchronization in Low and Medium Bit Rate Receivers

In a typical digital telecommunication system, the transmitter modulates groups of binary data (i.e. *symbols*) into phase, and/or frequency, and/or amplitude, differences of an analogue signal (*carrier*) [22]. *Symbol rate* is the ratio of the number of transmitted symbols per time unit. *Bit rate* is the ratio of the number of transmitted bits per time unit. In principle, for the receiver the symbol rate is known. However, the exact instance, that the modulated input signal must be sampled, is not specified. Furthermore, in the general case the receiver knows the width of the data bursts, but the starting po-

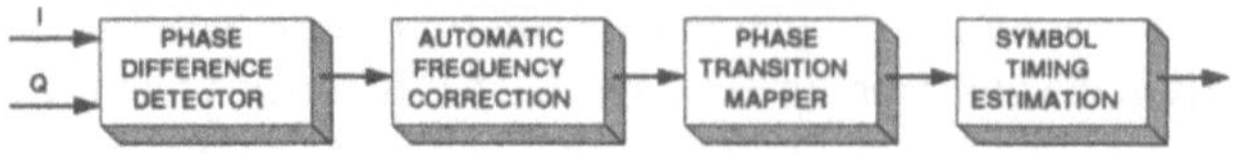

Figure 6.1 The Block Diagram of the DECT Receiving Algorithm

sition of them is not known. *Symbol timing synchronization* is the process of deriving at the receiver timing signals indicating where in time the transmitted signals are located [22, 10]. If symbol timing synchronization is not achieved, then even a small shifting in time of the sampling instances can result to receive erroneous data. *Frame synchronization* is the process of locating at the receiver the position of a synchronization pattern (marker), periodically inserted in the data stream by the transmitter [22, 10].

In the low (<100 kbits/s) and medium bit rate (<3 Mbits/s) communications domain, which covers a wide range of wireless and wire-line applications (including the popular GSM [11] and DECT [8]), the most common approach followed to overcome the synchronization problem is the use of *adaptive sampling through oversampling*. Implementations of wireless communication receivers that uses adaptive sampling through oversampling can be found in [19, 18, 17, 21, 25]. The wide use of this synchronization method is due to the fact that eliminates the need for a power hungry, analog VCO (Voltage-Controlled Oscillator) and provides both symbol timing and frame synchronization in one step [22].

Additionally, according to *adaptive sampling*, the sampling instance during the symbol period is selected separately for each data burst. *Oversampling* is the mechanism employed to choose the correct sampling instance during the symbol period for each data burst. Specifically, instead of sampling the analogue input signal once, the input signal is sampled N times during each symbol period. N is referenced as *oversampling ratio*. The input stream resulting from oversampling, is decomposed into N input data streams (one input stream per sampling instance during the symbol period) which are demodulated independently. For clarity reason, the streams that correspond to a fixed sampling instance during the symbol period are referred as sub-streams, while the stream acquired with oversampling is referred as the main stream. To identify the "correct" among the N input sub-streams, we search for an *a priori* specified synchronization pattern in each of them. The input sub-stream that corresponds to the correct sampling instance is the one that contains the synchronization pattern with an affordable deviation. In principle the synchronization pattern indicates the beginning of a frame. Thus, once this pattern is tracked down in one of the streams, it can be said that both symbol and frame synchronization has been achieved. After the reception of one frame, synchronization is assumed to be lost and the same procedure is repeated.

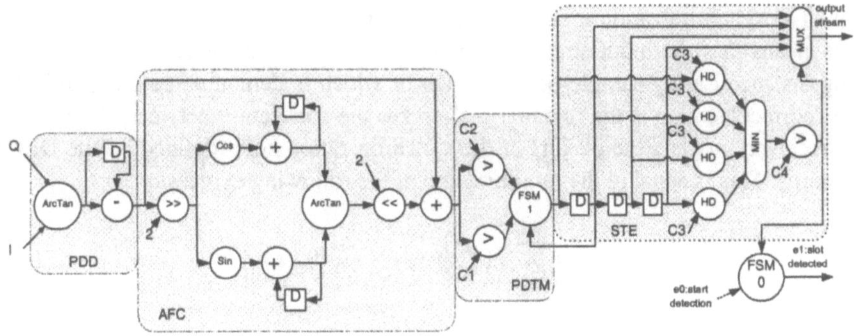

Figure 6.2 The Simplified CDFG of the DECT Receiving Algorithm

6.2.2 Main Demonstrator: The LPGD DECT Receiver

The main test-vehicle for both power management techniques presented in this chapter is a DECT receiver (Fig. 6.1). The algorithm developed for DECT receiving uses oversampling by 4 and consists of four main blocks:

- The Phase Difference Decoder Block (PDD)
- The Automatic Frequency Correction Block (AFC).
- The Phase Difference Transition Mapper Block (PDTM).
- The Symbol Timing Estimation Block (STE).

Specifically, the PDD block converts the I, Q, inputs, which represent Cartesian coordinates, to polar coordinates and calculates the angle differences between two successive symbols, as follows:

$$\Delta\varphi = \arctan \frac{Q}{I}\bigg|_t - \arctan \frac{Q}{I}\bigg|_{t-1} \tag{6.1}$$

The AFC block is responsible to correct errors (phase rotations) due to frequency drifts between receiver and transmitter oscillator. The phase rotation of the mth symbol caused from the local oscillator frequency drift is estimated using the following equation:

$$\varepsilon_\varphi = \frac{1}{4} \arctan \frac{\sum_{n=0}^{m} \sin\left(4\Delta\varphi(t - mT)\right)}{\sum_{n=0}^{m} \cos\left(4\Delta\varphi(t - mT)\right)} \tag{6.2}$$

where T is the period corresponding to the sample rate.

The error estimate is added to the phase difference computed by the PDD. The PDTM block maps the corrected phase difference to the appropriate binary symbol.

Finally the STE block is responsible for the symbol timing and frame synchronization. The synchronization method used is adaptive sampling through oversampling. That means that the input stream is decomposed to N streams (one for each sampling instance during sample period), each of which is demodulated independently. The "correct" stream is the one that maximizes the correlation between 32 consecutive received bits and the 32-bits of the synchronization field (defined by the DECT standard). This is equal to the minimization of the following expression:

$$C_i = \sum_{j=0}^{32} P_j \bigotimes S_{i,j}, \quad i \epsilon [0, N) \tag{6.3}$$

where: P_j are the bits of the synchronization sequence, the index i denotes one of the N streams that are derived by the decomposition of the input stream, and $S_{i,j}$ is the jth bit of the i stream. So, the above expression is computed for all the N input streams. If the minimum C_i is less than a threshold value then it is assumed that a frame has been detected, and during this frame the correct sampling instance is the one with the minimum C_i.

A simplified Control Data Flow Graph (CDFG) of the above described DECT receiving algorithm is illustrated in figure 6.2.

6.3 OPERATION/INSTRUCTION-LEVEL POWER MANAGEMENT

One of the most efficient low-power techniques applicable at all levels of abstraction of the design flow is the run-time power management. In the introduction of this chapter, prior existing runtime power management techniques have been briefly reviewed. The design of digital receiver subsystem circuits usually starts at the operation/instruction-level and if power management is postponed for the next design-levels, then power-management options can be disabled [29]. However, it seems that none of the prior existing techniques has addressed power management in this level.

In this chapter the methodology for the operation/instruction-level event-driven power management of digital receivers presented in [30] is analytically described. The methodology targets the exploration of the trade-off according to which on the one hand power consumption decreases because a resource is shutdown but on the other hand power consumption and area increases due to the presence of additional logic. In the general case, it is very difficult to explore this trade-off at the operation/instruction-level since the necessary timing and structural information are not available at this level of abstraction. However, the exploitation of specific characteristics of digital receiver has enabled the development of this methodology.

6.3.1 Power Model

The dynamic power dissipated by a CMOS combinational circuit can be expressed by:

$$Power = \sum_{i=1}^{N} C_{load_i} \cdot V_{DD}^2 \cdot f \cdot E_i \tag{6.4}$$

where C_{load_i} is the load capacitance at node i, V_{DD} is the power supply voltage, f is the frequency and E_i is the activity factor at node i.

The term $f \cdot E_i$ of equation 6.4 is is equal to the ratio of the number of node transitions from logic '1' to logic '0' divided by the total number of input vectors:

$$f \times E_i = \frac{N\_transitions_{1 \to 0_i}}{N\_vectors} \tag{6.5}$$

(6.4)$\bigwedge$(6.5)$\Rightarrow$

$$Power = \frac{V_{DD}^2}{N\_vectors} \cdot \sum_{i=1}^{N} C_{load_i} \cdot N\_transitions_{1 \to 0_i} \tag{6.6}$$

The power estimation problem is actually a two-dimensional problem, because both number of transitions and load capacitance have to be estimated. Additionally, operation/instruction-level power estimation has also to deal with the fact that the exact circuit structure is not yet fixed. The activity is estimated by calculating Hamming-distance at the output nodes of basic operations during functional simulation. For the effective capacitance estimation the models presented in [14] some of which are also available on the web [15] are used. Even though the accuracy of the power estimation model used here is poor, the claim is that it suffices for the purpose of comparing alternative power management scenarios at the operation/instruction-level. However, any other, operation/instruction-level power estimation method can be used for this purpose.

6.3.2 Proposed Methodology

Before any analysis, some definitions will be given for clarity reasons:

- *Unobservability* characterizes a fraction of a processor's behaviour and refers to the situation where the output values of this fraction are not relevant to the correct operation of the entire behaviour.

- *System period* is the minimum fraction of time during which a sequence of events is not repeated. In general, the system period is different from the period of the implicit outer loop of the Control/Data Flow Graph (CDFG). Actually, the system period is an integer multiple of the period of this implicit outer loop.

- *Event window* is a fraction of system period that lies between two events.

The proposed technique is based on the fact that at the operation / instruction-level unobservability may be introduced after the occurrence of an event. In all operation/instruction-level descriptions of receivers, computational clusters exist for which the goal is to check whether an event occurs or that the output variables are not modified between the occurrences of two events. Such clusters are characterized by unobservability for one or several event windows and their shutdown can lead to significant power savings. For example, a computational cluster responsible for synchronization does not change its outputs for a while after the synchronization is achieved. In an abstract operation/instruction-description (e.g. CDFG) it is not always clear whether a computational cluster performs useful computations or not. Thus a behavioural analysis is usually required in order to identify the computational clusters that can be shutdown and also the events that enable and disable them.

Step 1: Behavioral Analysis-Event Graph Formulation

Behavioral analysis indicates the candidate computational clusters for power management. This also involves the identification of the events that can trigger the shutdown of the computational clusters. The simplest way to perform behavioural analysis is simulation. However, simulation is not always needed, since behavioural analysis can also be performed manually by any designer familiar with the behavioural description of the design. In any case, the behavioural analysis can be visualized by the use of an *event graph*. This is a two dimensional graph, the horizontal dimension of which represents the system period and in which the events are arranged with respect to their occurrence sequence. This way the system period is divided into several event windows. In the vertical dimension the computational clusters are contained. If a computational cluster is observable in an event window then a line is drawn in the corresponding window. The absence of a line in an event window means that during this event window the corresponding computational cluster is unobservable.

A question that arises at this point is how the initial operation/instruction-level description is partitioned into computational clusters. The answer is simple; the event graph can be used in any level of granularity available. For example, a computational cluster can be a simple operation or a set of operations. Nevertheless, a fine-grained partitioning can result to a large exploration space that cannot be manually manipulated. On the other extreme, if large computational clusters are used for the purpose of power management exploration, then some power management opportunities may be lost. To overturn the partitioning problem, operations with common event windows, during which they are unobservable, can be grouped into a single computational cluster. Further grouping can be succeeded by ignoring event signals that are expected to occur in neighbouring time instances. However, the partitioning process can be avoided most of the times, since the operation/instruction-level description of a receiver algorithm is usually inherently partitioned into computational clusters (based on their functionality). The intuition is that this partitioning is in most cases optimal for the purpose of power management exploration.

Step 2: Power Management scenarios Extraction

The usage of all event signals for the purpose of power management is not necessary. A power management scenario that does not use all the event variables available can be more power efficient than the one that uses all the event variables. Any possible combination of the event variables that does not violate their initial ordering corresponds to a different power management scenario. If there are n events that introduce or cease unobservability for m computational clusters, then maximum $2^{n-1} - 1$ different power-management scenarios exist, each one corresponding to the use of a different set of events. Each power management scenario is characterized by different energy dissipation. This search space can become very large in cases that the number of events is large. However, in receiver applications the number of such events is small, permitting a manageable number of power management scenarios.

Step 3: Cost function evaluation

Each power-management scenario can be power-efficient or inefficient depending on the power consumed for the generation of the event variables and the power saved by the shutdown of the computational clusters. To decide whether a power-management scenario is efficient or not, the first step is to generate the logic that produces the event variables and insert them in the operation/instruction-level description of the algorithm. The generation of the event variables is not always needed since a variable that has identical behaviour with an event variable may already exist in the original description. In such cases it is assumed that no power overhead is introduced. Computational clusters that are unobservable during same event-window(s) can share shutdown variables. Whether a power management scenario can lead to power savings or not depends on two factors: First, the time fraction during which the computational clusters is shut down and their power consumption determines the amount of saved energy. Second, the possible additional logic needed for the event signals generation consumes additional power and time fraction during which this additional logic is needed to operate relates to the energy consumed by it. In order to decide whether a power-management scenario is efficient or not, the power of the original and transformed (i.e. original + power management related logic) operation/instruction-level description should be compared. The condition is simple; the power consumed by the transformed description must be less than the power consumed by the original description. A mathematical expression of the above statement follows:

$P_{Transf} < P_{Orig} \Rightarrow E_{Transf} \cdot T_{System\_Period} < E_{Orig} \cdot T_{System\_Period} \Leftrightarrow$

$$E_{Transf} < E_{Orig} \tag{6.7}$$

$$E_{Transf} = P_{Orig} \cdot T_{System\_Period} + P_{Add} \cdot T_{System\_Period} - \\ - \sum_i P_{Cluster_i} \cdot T_{Event\_Window_i} \tag{6.8}$$

$$E_{Orig} = P_{Orig} \cdot T_{System\_Period} \tag{6.9}$$

$(6.7) \wedge (6.8) \wedge (6.9) \Rightarrow$

$$P_{Add} \cdot T_{System\_Period} < \sum_i P_{Cluster_i} \cdot T_{Event\_Window_i} \tag{6.10}$$

where T_{System_Period} and T_{Event_Window} are the duration of system period and event-window respectively, $P_{Add.}$ is the power consumed by the event-variables generation logic and $P_{Cluster_i}$ is the power consumed by the computational cluster i.

Equation (6.10) denotes that power savings are achieved if the power consumed by the additional circuit (for the event variables generation) multiplied by the duration of systems period is less than the summation of the power saved by shutting down the computational clusters multiplied by the respective event-window duration. In other words, equation (6.10) describes the necessary condition for a power-management scenario to be valid (i.e. to be power efficient). A cost function that will help to choose among all valid power management scenarios can be derived directly from equation (6.10):

$$Cost\_Function = \frac{P_{Add} \cdot T_{System\_Period}}{\sum_i P_{Cluster_i} \cdot T_{Event\_Window_i}} \quad (6.11)$$

Unfortunately, both event window and system period duration are not known, because timing is fixed at a lower level of the design flow, during operation scheduling. But even if operation scheduling is performed, events do not always occur at a defined time instance. Therefore, in the general case the evaluation of the specific cost function is infeasible at this level, and can only be performed at the RT-level after the (custom) processor synthesis of both the alternative designs. It is feasible though to evaluate equation (6.11) at the operation/instruction-level in cases where time related information is part of the specifications. In this case the cost function must be evaluated for the worst case for each alternative power management scenario. The worst case for the cost function of equation (6.11) is the one that the shorter possible duration for the event windows and the longest duration of the system period are considered. When the duration of either an event window or the system period is not fragmented then their average duration can be fed to equation (6.11). In the receiver context such information is supplied by the telecommunication protocol, which is always part of the (if not whole the) specifications.

Almost always an area overhead is paid due to the insertion of logic for the generation of the event signals. It is assumed here that regardless of the amount of this overhead the choice is always to trade area for power. If this is not the case then the cost function used during operation/instruction-level power-management exploration must be modified in order to take into account area. Area estimation though is very difficult to be made at this level of abstraction since it strongly depends on the decisions made in the next design steps during operation scheduling, allocation and assignment.

Step 4: Implementation

Passing constraints from higher to lower levels of abstraction, is a key concept of the unified meta flow in chapter 2. Adherence to this concept, is also of critical importance for the efficient application of this methodology. Specifically, the methodology can be proven to be useless, if the decisions made by the operation/instruction-level power management exploration are not taken into account by the next step of the design process, which is the custom processor synthesis (operation scheduling-allocation-assignment). For example, if an operation contained in a computational cluster that was decided to be shutdown for a certain time fraction, and an operation outside this cluster are assigned to the same resource, then the shutdown of this resource either requires a complicated control logic (that consumes additional power) or is disabled. A custom-processor synthesis algorithm that targets functional pipelined architectures and that takes into account the power-management decisions made by the proposed methodology has been reported in [29].

6.3.3 Proposed Methodology Applied in a DECT Receiver

The above described methodology was applied for the design of the DECT receiver described in subsection 6.2.2. For the purpose of power management exploration, the

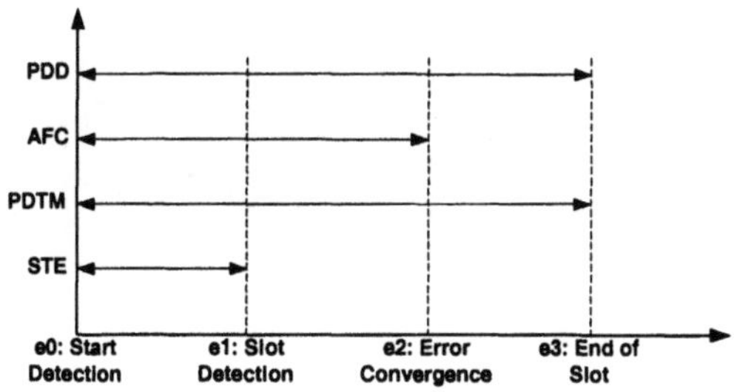

Figure 6.3 Event Graph for the DECT Receiver

computational clusters taken under consideration are PDD, AFC, PDTM and STE. The behavioural analysis indicates that the STE computational cluster does not need to operate after the header detection and up to receiving the whole slot. Also the AFC in order to estimate the error computes a series, which ones converges does not modify its output for the rest of the slot. The corresponding event graph is shown in Figure 6.3.

The events that divide the system period and determine the event windows are: the start detection ($e0$), the slot detection($e1$), the error convergence ($e2$) and the end of slot ($e3$). Event variables $e2$ and $e3$ are not present in the original CDFG. The additional logic for the generation of these event variables are shown in the transformed CDFG of Figure 6.4. For the event window between $e2$ and $e3$, the minimum duration (worst case), is equal to the three-quarters of the duration of a slot. The event window between $e1$ and $e3$ is always equal to the duration of a slot. The duration for system period is not fragmented but according to a probabilistic approach, the system period is on average 2 or 3 times the duration of the slot [20].

Three candidate power- management scenarios exist. According to the first one all the event variables ($e0$, $e1$, $e2$ and $e3$) are used. The second uses $e0$, $e1$ and $e3$, while the third uses $e0$, $e2$ and $e3$. In the first case, the STE computational cluster can be shutdown for the event-window between $e1$ and $e3$, AFC computational cluster can be shutdown for the event-window between $e2$ and $e3$, and we need the extra logic for the generation of both $e2$ and $e3$. In the second case, only the STE computational cluster can be shutdown, and obviously the additional logic that produces $e3$ is needed. In the last, the STE and the AFC computational clusters can be shutdown during the event window between $e2$ and $e3$ and the logic that produces the event variables $e2$ and $e3$ is needed.

All the candidate power management scenarios reduce the power budget. The cost function values and the power measures acquired after gate level estimation under real delay model, corresponding to each one of the above-described scenarios are illustrated in 6.5. The most efficient power-management scenario is the one that uses the variables $e0$, $e1$ and $e3$. This means that the presence of the logic that generates $e2$ introduces an energy overhead greater than the energy saved during the event window

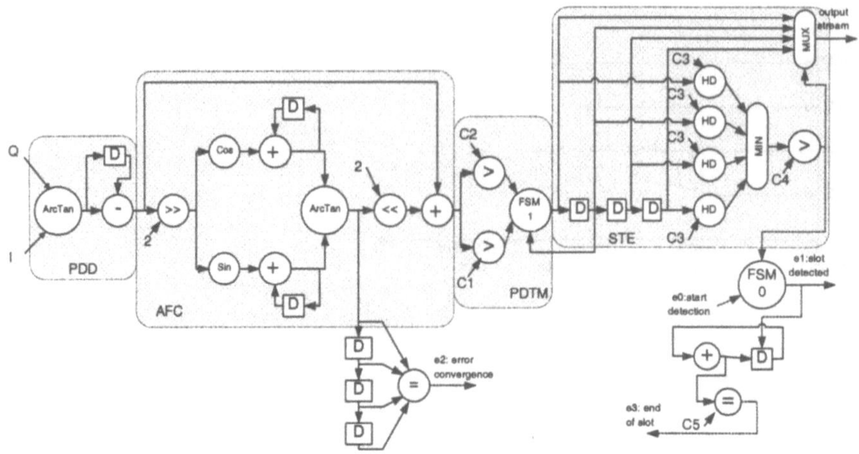

Figure 6.4 Transformed CDFG of the DECT Receiving Algorithm

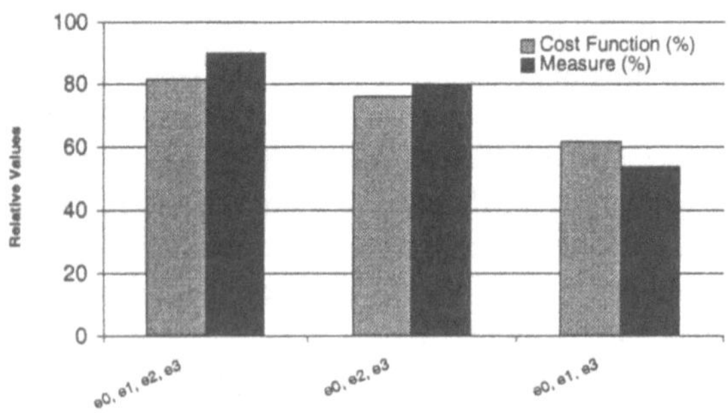

Figure 6.5 Power Management scenarios Comparison

between $e2$ and $e3$. This is due to the relatively small duration of the event window between $e2$ and $e3$.

6.4 DYNAMIC FREQUENCY SCALING

As previously commented, in the domain of low and medium bit rate receivers, the most common approach followed in order to overcome the synchronization bottleneck is the usage of adaptive sampling through oversampling. This approach generates at run-time power management opportunities, the exploitation of which can lead to significant gains in terms of energy dissipation. The following subsection underlines how energy can be saved in this class of receivers and explains why prior existing techniques fail to succeed.

6.4.1 Problem Definition and Related Work

Adaptive sampling through oversampling imposes that the main input stream acquired with oversampling is decomposed into N sub-streams, each of which must be demodulated separately. A block responsible for synchronization selects the "correct" among the N sub-streams. During the time intervals that synchronization is assumed to be achieved, the processing of the $N-1$ "wrong" sub-streams is useless. The power management discipline imposes that the unneeded operations should be avoided.

The application of this discipline in such a demodulation algorithm implies that the algorithm operates with different sample rates (i.e. $\times N$ and $\times 1$) for different time intervals (i.e. not synchronized and synchronized). In that sense, it can be said that the application of run-time power management, transforms the demodulation algorithm to a *multirate algorithm.*

The problem is to design *single rate* custom processor architectures for receivers, using adaptive sampling through oversampling, that implement the power management discipline by operating with different sample rates and that do not significantly increase area or critical path with respect to conventional architectures. By the term *single rate* we define custom processor architectures with no multiple or sub- clock lines, that would complicate routing and would cause clock skew problems.

Although extensive research work has been done towards custom processor architectures for DSP multirate algorithms [28, 7, 9, 24], the related techniques are not applicable in the case of demodulation algorithms. In the DSP community, a *multirate algorithm* is an algorithm with no constant sample rate throughout its description [28]. Typical examples of multirate algorithms are algorithms containing decimation and/or expansion. The main differences between receiving algorithms employing oversampling for synchronization purposes and typical multirate DSP algorithms, are the following:

- in the first case the whole algorithm operates in different sample rates for different time intervals, while in the latter case different parts of the algorithm operate with different sample rates and this does not changes over time.
- in the first case the algorithm description does not indicate its multirate nature, while in the latter case the multirate nature is inherent in the algorithm description (it is indicated by algorithm description entities such as decimation and expansion).
- in the first case the time instance at which the sample rate must change is not statically known and the related decision must be taken at run time, while in the latter case the sample rate does not change over time.

From the above, it is clear that existing techniques for mapping typical DSP multirate algorithms to single-rate custom processor architectures are not applicable for the purpose of power management for the class of receivers under consideration here.

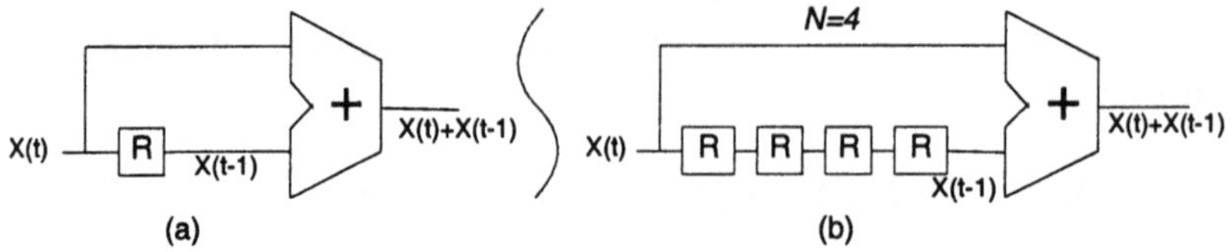

Figure 6.6 Example of Conventional Architecture Style

Furthermore circuit-level, run-time power management techniques, referenced in the Section 6.1 can be really helpful, but cannot be used for the purpose of forcing a receiver employing adaptive sampling through oversampling to operate only on the "correct" sample streams. It is almost straightforward, that in this case run-time power management is not relevant to the pre-computation idea [1], nor to the exploitation of FSMs *idle* states [2, 3], nor to the exploitation of observability *don't care* sets of specific computational blocks [27].

In [29], a run-time power management technique that satisfies the need for dynamic power management for this class of digital receivers is described. The following subsection describes the conventional architecture style for receivers employing oversampling for synchronization purposes, while the next subsection describes in a detailed manner the approach proposed in [29].

6.4.2 Conventional Architecture Style

Conventional approaches for receiver employing oversampling for synchronization purposes use a single clock line, with frequency equal to N times the symbol frequency, throughout the architecture including the analog-to-digital converter (ADC). Triggering the ADC N times during the symbol period results the ADC's output symbol stream (i.e. main stream) to contain multiplexed in time the symbol sub-streams (i.e. streams corresponding to fixed sampling instances during the symbol period). Each of the sub-streams must be demodulated separately, to finally decide which is the "correct" one. Area constraints and the related integration cost prohibit the use of parallel data-paths for the separate processing of the sub-streams. To succeed, resource reuse for all the sub-streams and perform the operations between samples belonging to the same sub-stream, N-stage shift-registers replace delay-registers. An example is given in Fig. 6.6: A part of a scheduled CDFG is shown on the left side and the corresponding hardware architecture for $N = 4$ is illustrated on the right side.

An abstract view of the conventional architecture style is illustrated in Fig. 6.7. It is assumed that a signal, which indicates to the receiver environment that synchronization is achieved, and one that indicates which is the correct among the sub-streams, are produced in the receiver. The first signal is referenced as TS while the second as $correct\_sample$. The output of the receiver can be in high impedance when TS indicates that there is no timing synchronization. When timing synchronization is es-

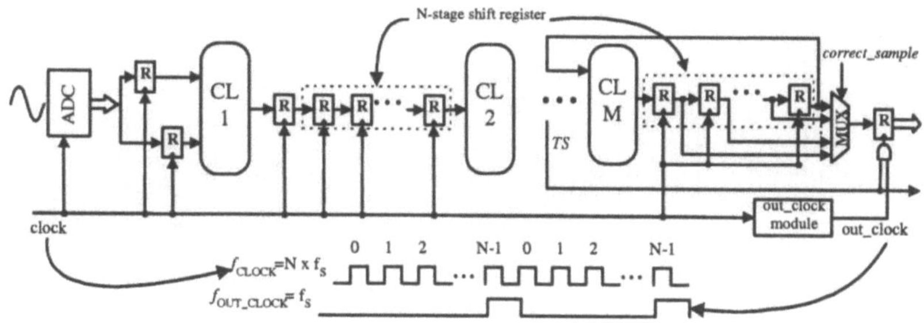

Figure 6.7 Conventional Architecture Style

tablished the *correct_sample* signal selects the stream that must be fed in the output and the output register is clocked at the symbol frequency.

6.4.3 Proposed Architecture Style

As mentioned earlier the aim here is to design custom processor architecture for receivers employing oversampling for synchronization purposes that avoid unnecessary computations during the time intervals that synchronization is achieved. For these intervals, the $N-1$ out of N computations are unnecessary. If during the synchronization intervals, frequency is down-scaled by a factor of N, then the number of performed computations equals the number of necessary computations. The number of computations performed per time unit is directly analogous to clock frequency. To identify the start and the end of the synchronization interval, the signal TS, described in subsection 6.4.2, can be used.

Although dynamic clock frequency scaling can be used to force the data-path to perform the number of necessary operations, still we need the operations to be performed on the correct data. To achieve this, firstly the registers that contain "useful" data must be fed to the inputs of the combinational logic blocks (CL), and secondly registers that contain "useless" data must be manipulated in such a way that they will not affect the correct operation of the receiver.

An architecture template, which implements an effective strategy of steering registers' clock gating and bypass mechanisms that are necessary for run-time power management, is illustrated in Fig. 6.8. The main modifications with respect to the conventional architecture model (Fig. 6.7), as far as the architecture building blocks are concerned, are the following:

1. All registers (R), of the conventional architecture, are replaced, in the proposed architecture, with registers (R') the output of which is either the stored value or their input with respect to the value of a control signal.

2. The *clock frequency control* block is added. This block is responsible to divide the clock frequency for the intervals that synchronization is achieved. For these

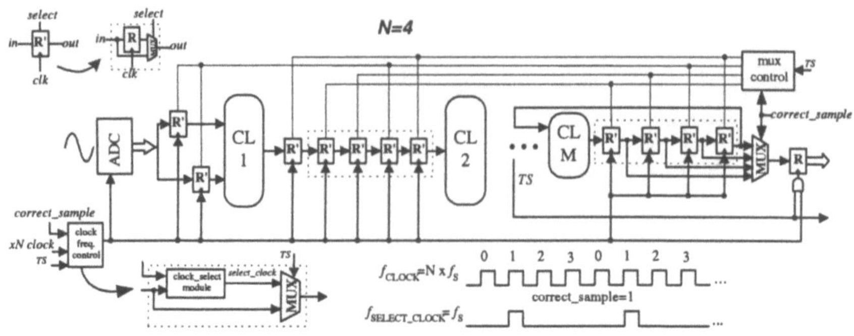

Figure 6.8 Proposed Architecture Style

intervals the clock must have the appropriate waveform to force the ADC to sample on the correct sampling instance during the symbol period. So, signal TS triggers the frequency scaling and the *correct_sample* signal is used to determine the down-scaled clock waveform.

3. The *mux_control* block is added. This block is responsible to control the multiplexers that bypass the registers. The registers that must be bypassed are those that contain "wrong" sub-streams samples at the time instance that the clock frequency is scaled form $\times N$ down to $\times 1$ the sample frequency. The same registers are bypassed during all the synchronization interval. It is assumed that when the signal TS is activated, the register that is closer to the output contains a sample from the Nth stream. Thus, if the Mth sub-stream is the correct, then the Mth closer register to the output must not be bypassed. Additionally, all the registers located at a distance of a multiple of N from that register must not be bypassed. The rest of the registers is bypassed. To achieve that, the output of the *mux_control* block is an N-bits word. The first bit of this word is connected to the registers which distance from the register that is closer to the output, is a multiple of N. The second bit of this word is connected to the registers which distance from the 2nd most closer register to the output, is a multiple of N, and so on. If the Mth stream is the correct, then during synchronization the *mux_control* block feeds a logical "1" at the Mth bit and logical "0" to the rest bits of its output word.

In this way, for the time intervals that synchronization is not achieved, the $\times N$ frequency clock is fed to the architecture and none of the registers is bypassed. Thus for these time intervals the operation is identical to this of the conventional architecture. For the intervals that synchronization is achieved, clock frequency is down scaled by a factor of N and the registers that contain the "wrong" data are bypassed, forcing the receiver to operate only on the correct data. Taking into account that a large amount of unneeded operations is avoided with the proposed architecture, it can be inferred

N_Taps	N	Conventional	Proposed	Dif. %
4	4	10.405	3.320	-68.09
4	8	21.549	6.502	-69.83
4	16	47.460	12.937	-72.74
8	4	27.032	8.478	-68.64
8	8	54.945	16.630	-69.73
8	16	116.328	33.166	-71.49

Table 6.1 Power in (mW) for $T_S/T_F = 1/128$

that this architecture will consume significantly less power compared to the conventional one. Specifically, our experimental results point out that power savings range from 40% up to 70%, depending on the selected oversampling ratio and the receiver structure.

6.4.3.1 Side effects. Although the power savings are expected to be significant, they do not come at no cost. Specifically, two main design parameters, namely area and critical path, are negatively effected with the proposed architecture when compared to the conventional one. Area is increased due to the additional multiplexers at the output of the registers, due to the related control logic and interconnection and due to the insertion of the clock frequency scaling block. Critical path is increased by the delay of a multiplexer, when operating at $\times N$ the symbol frequency. When operating at $\times 1$ symbol frequency the critical path is multiplied by a factor of N. However, if there no timing violations when operating at $\times N$ the symbol frequency, there will be no timing violations when operating at $\times 1$ the symbol frequency. This is because, in the latter case both critical path and clock period are multiplied by the same factor (N). Although the expansion of the critical path during synchronization intervals does not cause extra timing problems, it indirectly affects power consumption, since in this case the spurious transitions are propagated in a longer path. This may decrease, but cannot overturn the expected power savings.

6.4.4 Experimental results

One of the main parts of receivers is the digital filtering stage. For this reason, Finite Impulse Response (FIR) filters are chosen as demonstrator application. Four- and eight-taps FIR filters are implemented. A low-power implementation of the FIR filters is considered, where multiplications are replaced with shift-add operations. The filter coefficient bit-width is 10 bits, while the symbol bit-width is 9 bits. For each FIR filter three different oversampling ratios are considered, namely 4, 8 and 16. The frame is considered to consist of 128 symbols. Finally, four different values for the width of the synchronization pattern are considered, namely 1, 2, 4 and 8 symbols. For each different FIR filter configuration, two implementations are considered: one according to the conventional architecture and one according to the proposed architecture. For the FIR implementation, the SYNOPSYS and Mentor-Graphics CAD tools were em-

N_Taps	N	Conventional	Proposed	Dif. %
4	4	10.403	3.354	-67.76
4	8	21.322	6.539	-69.33
4	16	47.240	12.929	-72.63
8	4	26.997	8.569	-68.26
8	8	54.458	16.747	-69.25
8	16	115.754	33.115	-71.39

Table 6.2 Power in (mW) for $T_S/T_F = 2/128$

ployed and the 0.6 micron AMS cell-library was used for mapping. For each different FIR filter the power consumption, the area and the critical path were measured. Power measurements were acquired with toggle-count during logic-level simulation under real delay model and capacitance estimates provided by the CAD tools, assuming 5V power supply and 4K highly-correlated, 9-bit, valid input vectors. The latter means that if the oversampling ratio is N then the number of input vectors used is $4000 \times N$. The $4000 \times (N - 1)$ vectors that correspond to the wrong sampling instances were generated with a random variation from the corresponding valid vector in the range of $\pm 10\%$. For area and critical path measurements, the reports provided by the CAD tools were used.

Table 6.1 - 6.4 illustrate the power consumed for both the original and proposed implementations for four different cases. Each case corresponds to a different ratio T_S/T_F of synchronization pattern width to frame width. The ratio T_S/T_F determines the relation between the sizes of the time intervals operating at oversampling and symbol frequency. It was expected that as the latter ratio increases, power savings decrease. Experimental results indicate that the effect of the ratio T_S/T_F on power savings is weak. For example, the power savings for the case that $T_S/T_F = 1$ are on average 1.57% larger than the power savings for the case that $T_S/T_F = 8$.

Additionally, from Tables 6.1 - 6.4, it can be observed that the amount of power saved by the proposed technique, mainly depends on the oversampling ratio. This is logical, since the oversampling ratio determines the difference between the power dissipated for the time intervals operating at oversampling and symbol frequency. Furthermore the percentage of power savings does not seem to depend on the design size since the results are very close for both 4-taps and 8-taps FIRs. In any case, the proposed architecture model consumes significantly less power (on average 69.43%) than the original architecture model.

Table 6.5 and 6.6 illustrate the area and critical path measures respectively, for both the original and proposed implementations. It can be observed that the area overhead introduced by the multiplexers, the additional control logic and interconnections of the proposed architecture, is on average 1.96%. Furthermore, the proposed architecture increases the critical path of the original design by the delay of the multiplexer that is added at the output of the registers. This increase is less than 0.5 ns in any case. The overheads in area and critical path introduced by the proposed technique are

N_Taps	N	Conventional	Proposed	Dif. %
4	4	10.287	3.381	-67.13
4	8	21.101	6.506	-69.17
4	16	46.647	12.919	-72.30
8	4	26.735	8.722	-67.38
8	8	53.869	16.709	-68.98
8	16	114.308	33.019	-71.11

Table 6.3 Power in (mW) for $T_S/T_F = 4/128$

N_Taps	N	Conventional	Proposed	Dif. %
4	4	10.043	3.391	-66.24
4	8	20.572	6.532	-68.25
4	16	45.432	12.882	-71.65
8	4	26.064	8.764	-66.38
8	8	52.472	16.739	-68.10
8	16	111.204	32.793	-70.51

Table 6.4 Power in (mW) for $T_S/T_F = 8/128$

N_Taps	N	Conventional	Proposed	Dif. %
4	4	5344.91	5456.64	2.09
4	8	5957.43	6058.57	1.70
4	16	7181.44	7270.18	1.24
8	4	10802.74	11011.79	1.94
8	8	11832.08	12022.89	1.61
8	16	13895.81	14602.52	3.19

Table 6.5 Area measures in $mils^2$

considered to be negligible compared to the corresponding power savings. It must be stressed here, that the latency of the design is not affected by the proposed technique.

Finally, an interesting observation is the following: the power consumption of the original case when $N = 4$ and 8 is greater than the power consumption of the proposed case when $N = 8$ and 16, respectively. This means that the application of the proposed technique enables the use of a higher oversampling ratio (which means higher reception quality), while not exceeding the initial power budget.

Additionally, the DECT receiver described in subsection 6.2.2 and the corresponding FIR filter [21] were implemented according to the conventional and proposed architecture style. For this DECT receiver the oversampling ratio is 4. The DECT slot has 480 bits with 32 bits synchronization pattern. The two architectures were mapped

N_Taps	N	Conventional	Proposed	Dif. %
4	4, 8, 16	33.60	34.04	1.31
8	4, 8, 16	38.76	39.20	1.14

Table 6.6 Critical path measures ($nsec$)

	Conventional	Proposed	Dif. %
Current (μA)	284	170	-40.1
Gate Count	42433	47765	12.56

Table 6.7 Measurements for the FPGA implementation of the DECT receiver

in a XILINX xc4062xla FPGA, package hq240. Current measurements using 4000 samples from encoded voice were carried out. As shown from Table 6.7, the proposed architecture consumes 40.1% less current and has 12.56% greater gate count compared to the conventional one.

Although power gains are significant, it is expected that in an ASIC implementation power gains would be even greater. This is because in an FPGA implementation the interconnect power dominates on the total power budget [13], while in an ASIC implementation this is not yet true. So, in an ASIC implementation the additional interconnections of the proposed architecture style, would add a smaller power overhead.

6.5 CONCLUSIONS

Two run-time power management techniques applicable in a wide class of telecommunication receivers are presented. First, a operation/instruction-level event-driven power management methodology is described. This methodology targets the exploration of the trade-off according to which on the one hand power is saved by shutting-down parts of the circuit but on the other hand power is increased by the additional logic required. Second, a custom processor architecture style that employs dynamic frequency scaling for the purpose of power management, is described. The adaptation of this architecture style introduces significant power savings, while increasing negligible area and critical path. The efficiency of these techniques is proven by their application in a real-life DECT receiver.

References

[1] M. Aldina, J. Monteiro, S. Devadas, A. Ghosh, M. Papaefthymiou, "Precomputation-Based Sequential Logic Optimization for Low Power", *IEEE Transactions on Very Large Scale Integration Systems*, vol. 2, no. 4, pp. 426-436, 1994.

[2] L. Benini, P. Siegel, G. De Micheli, "Automatic Synthesis of Gated Clocks for Power Reduction in Sequential Circuits", *IEEE Design and Test of Computers*, vol. 11, no. 4, pp. 32-40, 1994.

[3] L. Benini, G. De Micheli, "Transformation and Synthesis of FSMs for Low Power Gated Clock Implementation", *IEEE Transaction on Computer Aided Design*, vol. 15, no. 6, pp. 630-643, 1996.

[4] L. Benini, G. De Micheli, *DYNAMIC POWER MANAGEMENT: Design Techniques and CAD tools*, Kluwer Academic Publishers, 1998.

[5] L. Benini and G. De Micheli, "System-Level Power Optimization: Techniques and Tools", in *proc. ACM/IEEE International Symposium on Low Power Electronics & Design*, pp. 288-293, Aug. 1999.

[6] L. Benini, A. Bogliolo, G. Paleologo and G. De Micheli, "Policy Optimization for Dynamic Power Management", *IEEE Transactions on Computer-Aided Design of Integrated Circuits and Systems*, vol. 18, no. 6, pp. 813-833, June 1999.

[7] S. S. Bhattacharyya, and A. E. Lee, "Memory management for data flow programming of multirate signal processing algorithms", *IEEE Transactions on Signal Processing*, vol. 42, pp. 1190-1201, May 1994.

[8] ETS 300 175 : "*Radio Equipment and Systems (RES); Digital European Cordless Telecommunications (DECT); Common Interface (CI)*".

[9] T. C. Denk and K. K. Parhi, "Synthesis of Folded Pipelined Architectures for Multirate DSP Algorithms", *IEEE Transactions on Very Large Scale Integration Systems*, Vol. 6, No. 4, pp. 595-607, Dec. 1998.

[10] J. D. Gibson, *The communications handbook*, CRC Press and IEEE Press, 1997.

[11] GSM 05.05 (ETS 300 577) "*European digital cellular telecommunications system (Phase 2); Radio transmission and reception*".

[12] C.-H Hwang, A. C.-H Wu, "A Predictive System Shutdown Method for Energy Savings of Event Driven Computation", in *proc. IEEE/ACM International Conference on Computer Aided Design*, pp. 28-32, San Jose, CA, Nov. 1997.

[13] E. Kusse and J. Rabaey, "Low-energy embedded FPGA structures", in *proc. ACM/IEEE International Symposium on Low Power Design*, pp. 155-160, San Diego, CA, Aug. 1998.

[14] P. Landman, "Low power architectural design methodologies", Ph.D. Dissertation, U.C. Berkeley, Aug. 1994.

[15] D. Lidsky and J. Rabaey, "Early exploration - A World Wide Web application", in *proc. ACM/IEEE Design Automation Conference*, pp. 27-32, June 1996.

[16] E. Macii, M. Pedram, F. Somenzi, "High-Level Power Modeling, Estimation, and Optimization", *IEEE Transactions on Computer-Aided Design of Integrated Circuits and Systems*, Vol. 17, No. 11, pp. 1061-1079, Nov. 1998.

[17] E. Metaxakis, A. Tzimas and G. Kalivas, "A low complexity baseband receiver for direct conversion DECT-based portable communications", in *proc. IEEE International Conference on Universal Personal Communications*, pp. 45-49, Florence, Italy, 1998.

[18] L. Molisch, B. Lopez, M. Paier, J. Fuhl, E. Bonek, "Error Floor of Unequalized Wireless Personal Communications Systems with MSK Modulation and Training-Sequence-Based Adaptive Sampling", *IEEE Transactions on Communications*, Vol 45 No. 5, pp 554-562, May 1997.

[19] K. Murota, K. Hirade, "GMSK Modulation for Digital Mobile Radio Telephony", *IEEE Transactions on Communications*, Vol. Com 29, No 7, pp. 1044-1050, July, 1981.

[20] M. Perakis, D. Soudris, C. Goutis and A. Thanailakis, "Low-Power Architecture Design of GSM/DECT Algorithms", deliverable report LPGD\WP2\DUTH\D3.3R1, Nov. 1998.

[21] M. Perakis, et al "The VLSI Implementation of a Baseband Receiver For DECT-Based Portable Applications", in *proc. IEEE International Symposium on Circuits and Systems*, Vol. I, pp. 198-201, Orlando, Florida, USA, 1999, .

[22] J. G. Proakis, *Digital Communications*, 3rd edition, McGraw-Hill, New York, NY 1995.

[23] Q. Quiu and M. Pedram, "Dynamic Power Management Based on Continuous-Time Markov Decision Processes", in *proc. ACM/IEEE Design Automation Conference*, pp. 555-561, June 1999.

[24] S. Ramanathan, V. Visvanathan, S. K. Nandy, "A computational engine for multirate FIR digital filtering", *Signal Processing*, Elsevier Science, Vol. 79, Issue 2, pp. 213-222, Dec. 1999.

[25] D. Soudris et al, "A Low-Power Design of a DCS1800-GSM/DECT Modulator/Demodulator", to be published in *proc. IEEE International Symposium on Circuits and Systems*, Geneva, Switzerland, May 2000.

[26] M. B. Srivastava, A. P. Chandrakasan, R. W. Brodersen, "Predictive System Shutdown and other Architectural Techniques for Energy Efficient Programmable Computation", *IEEE Transactions on Very Large Scale Integration Systems*, Vol. 4, No. 1, pp. 42-55, March 1996.

[27] V. Tiwari, S. Malik, P. Ashar, "Guarded Evaluation: Pushing Power Management in Logic Synthesis/Design", in *proc. ACM/IEEE International Symposium on Low Power Design*, pp. 221-226, Dana-Point, CA, April 1995.

[28] P. P. Vaidyanathan, *Multirate Systems and Filter Banks*, Englewood Cliffs, NJ: Prentice Hall, 1993.

[29] N. D. Zervas, D. Soudris, C. E. Goutis and A. Thanailakis, "Low Power Methodology for Transformations of Wireless Communication Algorithms", deliverable report LPGD\WP2\DUTH\D2.2R1, Jan. 1999.

[30] N. D. Zervas, D. Soudris, S. Theoharis, C. E. Goutis and A. Thanailakis, "A Methodology for the Behavioral-Level Event-Driven Power Management of Digital Receivers", to be published in *proc. IEEE International Symposium on Circuits and Systems*, Geneva, Switzerland, May 2000.

7 POWER EFFICIENT SYNTHESIS OF SUM-OF-PRODUCTS COMPUTATIONS

Konstantinos Masselos and Costas E. Goutis

Department of Electrical and Computer Engineering
University of Patras, Greece

masselos@ee.upatras.gr,goutis@ee.upatras.gr

Sum-of-products computations are widely used in multimedia and communications systems. Techniques for the power efficient data path synthesis of sum-of-products computations are presented in this chapter. Simple and efficient heuristics for the instruction-level scheduling and assignment steps are described. These steps are crucial sub-steps of the custom processor synthesis stage of the system level design meta-flow proposed in chapter 2. The proposed heuristics exploit the inherent independence of the sum-of-products computations to formulate the synthesis tasks using the concept of the Traveling Salesman's Problem. In this way the synthesis tasks can be solved very efficiently. Different partly static cost functions are proposed to drive the synthesis tasks. The proposed cost functions target the power consumption either in the buses connecting the functional units with the storage elements or in the functional units. The partly static nature of the proposed cost functions reduces the time of the synthesis procedure. Experimental results from different relevant digital signal processing algorithmic kernels prove that the proposed synthesis techniques lead to significant power savings.

7.1 INTRODUCTION

Portability as well as packaging and cooling issues made power consumption an important design consideration. The recent advances in the areas of communications and multimedia technology made available a large number of complex power hungry systems such as cellular phones, pagers, modems, video phones and digital video cameras. Such systems may operate in portable environments. All these systems make extensive use of digital signal processing (DSP). Since power consumption became an important design issue especially for the development of portable systems, low power digital signal processing became an increasingly important research area.

F. Catthoor (ed.),
Unified Low-power Design Flow for Data-dominated Multi-media and Telecom Applications, 159-180.

Optimization for power can be achieved in all the levels of the design hierarchy. The switching component of power dissipation can be optimized in the high levels of abstraction where the most significant power savings can be achieved [18]. A large number of custom data path synthesis techniques for power optimization have been proposed [3], [5], [7], [14], [19], [20], [22]. Techniques that improve the regularity between flow graph clusters in order to reduce the sharing overhead for clusters that share the same resource have been proposed in [9], [13]. Such techniques are based on the application of algebraic transformations that reduce the operator cost. The techniques described in [9] improve power consumption implicitly while the techniques described in [13] explicitly aim at reducing power consumption and mainly target address expressions. A large number of techniques and methodologies for low power digital signal processing have been proposed as well. Power optimization techniques for linear computations in digital signal processing algorithms are described in [2], [15]. A technique for low power digital signal processing in a programmable processor context is described in [10]. Several techniques for power optimization of FIR filters have been proposed in [6], [8], [12], [17], [21]. A detailed comparison of the techniques described in this chapter with the above mentioned existing approaches is included in 7.6

One of the most common operations performed in digital signal processing applications is the sum-of-products computation. The power consumed in sum-of-products computations form an important part of the total power budget of a DSP system [11]. Two main categories of DSP algorithmic kernels requiring sum-of-products computations are:
a) Convolutional algorithms
The basic computation of the algorithms of this category is the convolution between data and coefficient vectors and is described by equation 7.1. A typical algorithm belonging to this category is the FIR filtering.

$$Y_n = \sum_{i=0}^{N-1} C_i \cdot X_{n-i} \tag{7.1}$$

where $C_i^{\prime}s$ are the coefficients of the filter and X_n, Y_n are the n_{th} terms of the input and output sequences respectively.
b)Transformational algorithms
The basic computation of these algorithms is the matrix-vector multiplication between a coefficient matrix and a data vector and is described by equation 7.2. Typical examples of algorithms belonging to this category are the common DSP transformations (like DCT and DFT-FFT).

$$\begin{bmatrix} Y_0 \\ Y_1 \\ \vdots \\ Y_N - 1 \end{bmatrix} = \begin{bmatrix} C_{00} & \cdots & C_{0(M-1)} \\ C_{10} & \cdots & C_{1(M-1)} \\ \vdots & & \vdots \\ C_{(N-1)0} & \cdots & C_{(N-1)(M-1)} \end{bmatrix} \times \begin{bmatrix} X_0 \\ X_1 \\ \vdots \\ X_{M-1} \end{bmatrix} \tag{7.2}$$

where Y is the N-point output data vector ($Y_i^{\prime}s$), C is the $N \times M$ coefficient matrix ($C_{ij}^{\prime}s$) and X is the N-point input data vector ($X_i^{\prime}s$). For the algorithms of this category there are two options: I) Sequential output computation and II) Parallel output computation.

In this chapter novel techniques for low power data path synthesis of sum-of-products computations are proposed. The proposed techniques exploit two basic characteristics of sum-of-products computations between data and coefficients: a) The fact that the partial products of the sum-of-products structures are basically independent. b) The static nature of the coefficients. Simple and efficient instruction-level scheduling and assignment heuristics for independent computations are proposed. Different cost functions are proposed targeting the power consumption both in the buses connecting the storage elements to the functional units and in the functional units. The partly static nature of the proposed cost functions leads to reduction of the time of the synthesis procedure. The proposed data path synthesis techniques can be included in the custom processor synthesis step of the system level design meta-flow proposed in chapter 2.

The rest of the chapter is organized as follows: The target architecture model is described in section 7.2. The proposed synthesis techniques are described in detail in section 7.3. Power related cost functions are proposed in section 7.4. Experimental results for several digital signal processing algorithms are presented in section 7.5. A comparison of the proposed synthesis techniques with existing relevant approaches is attempted in section 7.6. Finally conclusions are offered in section 7.7.

7.2 TARGET ARCHITECTURE

The target architecture model is shown in figure 7.1. The foreground registers store the data terms required for the computation of each output term. In the case of original background data storage the use of registers allows exploitation of the data reuse present in both convolutional and transformational basic computations, favoring power consumption reduction. The coefficients are usually stored in a separate memory. The computations are performed in multiply/accumulator based data paths. Word serial implementations of the functional units are assumed. It is also assumed that for each functional unit used for the computation there are two different buses one for the data and one for the coefficients connecting the corresponding storage elements to the functional unit.

The number of functional units used for the computation is determined by the performance requirements of the target application. When multiple functional units are present, two main options exist for the storage and interconnect organization and they are described in figures 7.2 and 7.3. In the first case, data and coefficients are stored in central storage elements. These storage elements are multi-port and connected through multiple buses to the functional units. The second architecture is more local. Each functional unit has local data and coefficient storage elements. These memories are single-port and connected through a single local bus to the corresponding functional unit. The proposed synthesis techniques can be applied in both cases and even in combinations of them.

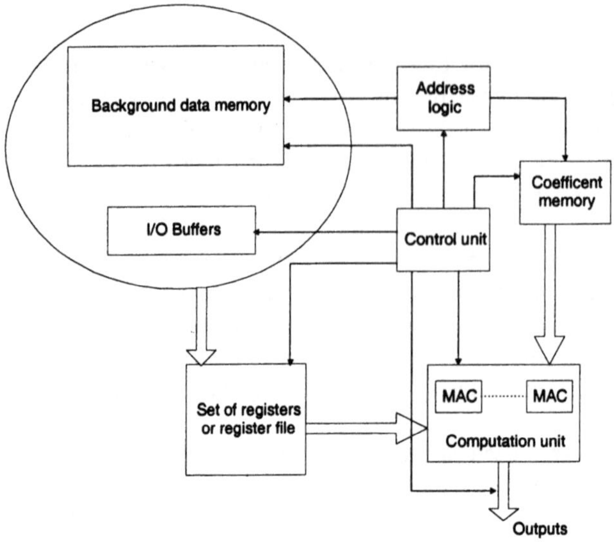

Figure 7.1 Target architecture model

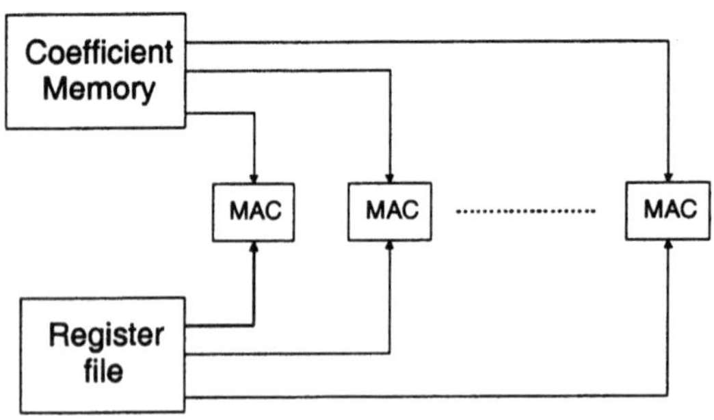

Figure 7.2 Centralized memory and bus organization

7.3 PROPOSED SYNTHESIS TECHNIQUES

The proposed synthesis techniques are described in this section. Synthesis techniques for independent data path operations are described first. Then these techniques are adapted for the synthesis of sum-of-products computations.

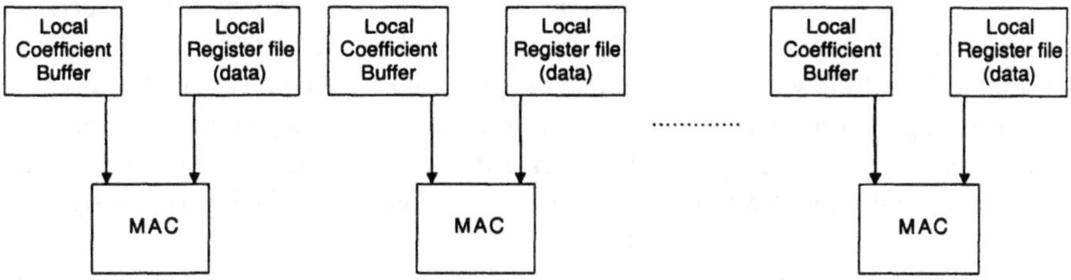

Figure 7.3 Localized memory and bus organization

7.3.1 Synthesis Techniques for Independent Data Path Operations

In this section data path synthesis techniques for sets of independent data path operations of the same type are proposed. The independence of the operations means that there are no precedence or ordering constraints related to the execution of the operations.

Let assume an algorithm with A independent operations of the same type. The synthesis of these operations on a single functional unit requires only the scheduling of the operations. For this case the following approach is proposed: A power related cost $PC_{i,j}$ is assigned to each pair (i,j) of operations. This cost is related to the power consumed when operations i, j are successively executed on the same functional unit and it is used to drive the synthesis procedure. In the next step the graph of the operations is built with the vertices representing the operations and the edges the unconstrained transitions between operations. Since no dependencies exist between operations the graph is complete (all pairs of operations are connected through an edge). To each edge (i,j) the cost $PC_{i,j}$ corresponding to the operations i, j connected by the edge is assigned. If $PC_{i,j} = PC_{j,i}$ the graph is undirected while if $PC_{i,j} \neq PC_{j,i}$ the graph is directed. The power efficient instruction-level scheduling problem under the above assumptions can be formulated as a Traveling Salesman's Problem (TSP) on the graph described above. To derive a valid schedule it is required to pass through all vertices (operations) only once and then return to the first vertex for the next algorithm's iteration. The cost function that is minimized during the scheduling procedure is the total power cost of the schedule and is given by the equation:

$$Cost\_function\_1 = \sum_{i=0}^{A-1} PC_{i,i+1} + PC_{A-1,0} \tag{7.3}$$

where the last term $PC_{A-1,0}$ represents the inter-iteration power cost i.e. the power consumed during the transition from the last operation of one algorithm's iteration to the first operation of the successive algorithm's iteration.

In the case where the A operations are distributed to a number B of functional units and assuming that A/B operations are assigned to each one of the B functional units (for load balancing reasons) the synthesis procedure will include two steps:

a) Assignment of operations to functional units: The A operations are grouped to B clusters of A/B operations. Each cluster is assigned to one of the B functional units available. The following simple heuristic is proposed for the power-conscious grouping of the operations: The A operations are scheduled assuming single functional unit implementation using the procedure described above. The sequence of operations is then sequentially divided to B parts of A/B operations each. Each part is assigned to one of the available functional units.
b) Scheduling of the operations assigned to each functional unit: The A/B operations assigned to each of the functional units are scheduled in a second step. The power conscious single functional unit instruction-level scheduling procedure described earlier is used for this reason.
The cost function that is optimized by the two step synthesis procedure described above is the total power cost of the complete computation over the different functional units and is given by the equation:

$$Cost\_function\_B = \sum_{k=0}^{B-1}\left(\sum_{l=0}^{(A/B)-1} PC_{k(l),k(l+1)} + PC_{k(A/B-1),k(0)}\right) \tag{7.4}$$

where k(l) is the operation scheduled l_{th} on the functional unit k.

Several algorithms have been proposed for the solution of the TSP problem. If the size of the problem is relatively small, an exact solution can be found in short time. A large number of exact algorithms have been proposed for the TSP which can be best understood and explained in the context of Integer Linear Programming (ILP) [16]. The NP-complete class of the problem motivated the research for heuristic algorithms. Christofides in [4] proposed a heuristic using a minimum-cost matching algorithm and requiring $O(n^4)$ time to find a nearly optimal solution. From the above it becomes clear that the proposed TSP based synthesis heuristics offer the advantage of finding a good quality solution (in some cases exact) in short time. In the prototype implementation of the proposed synthesis techniques an exact algorithm is used for the solution of the TSP problem when the number of the graph nodes is smaller than 12. When the number of nodes is larger than or equal to 12 a fast non-exact heuristic based on genetic algorithms is used.

7.3.2 Synthesis of Sum-of-Products Computations

Sum-of-products computation is a typical example of computational kernel consisting of independent data path operations. In this section the synthesis techniques described in section 7.3.1 are adapted for the synthesis of the two types of sum-of-products DSP kernels described in section 7.1

7.3.2.1 Single functional unit case. In this section realizations of sum-of-products computations on single functional units are assumed. The instruction-level scheduling problem is formulated for the different categories of DSP kernels.

Convolutional algorithms

The convolution operation performed by the algorithms of this category for the evaluation of an output term is described by the following equation:

$$y_n = \sum_{f(k)=0}^{N-1} C_{f(k)} \cdot X_{n-f(k)} \tag{7.5}$$

where y_n is the n_{th} term of the output sequence, C is the coefficient vector and X is the vector of the input data. The computation is performed according to the scheduling function $f(k)$, k=0,...,N-1. The scheduling function generates for each partial product k a relative order of computation and it is one-by-one and onto. In the original case $f(k) = k$. The total power cost (PC) of the sequence of evaluation of the partial products on a single functional unit as determined by the scheduling function f is given by the following equation:

$$PC(f) = \sum_{f(k)=0}^{N-1} PC(p_{f(k)}, p_{f(k)+1}) + PC(p_0, p_{N-1}) \tag{7.6}$$

where $p_{f(k)}$ is the partial product $c_{f(k)} x_{n-f(k)}$. The term $PC(p_0, p_{N-1})$ denotes the power cost between the first ($f(k) = 0$) and the last ($f(k) = N-1$) partial products (according to the ordering function $f(k)$). This term is included because after the computation of the last partial product (p_{N-1}) for the output term y_n the first partial product (p_0) for the term y_{n+1} must be computed (inter iteration factor). The power cost for a pair of partial products $PC(p_k, p_l)$ is given by the power related cost function selected to drive the synthesis procedure. The aim of the proposed synthesis techniques is to derive the ordering function f that minimizes the total power cost of the convolutional computation as given by equation 7.6.

The instruction-level scheduling problem for the convolutional algorithms is formulated as a Traveling Salesman's Problem (TSP). The graph $G(V, E)$ of the problem consists of the set V of the N vertices, which are the partial products required for the computation of an output term, and the set E of edges which model the unconstrained transitions between partial products. The problem's graph is complete i.e. each vertex pair is connected by an edge. To each edge of the graph the power cost PC between the two partial products that the edge connects is assigned. A closed path over all vertices (partial products) must be found, without passing from a vertex more than once, resulting to a minimum power cost.

Transformational algorithms - Sequential output computation

The sequential computation of the output terms of a transformation of length $N \times M$ is described by the equation:

$$y_{f(i)} = \sum_{g(i,j)=0}^{M-1} c_{f(i),g(i,j)} \cdot x_{g(i,j)} \tag{7.7}$$

where y's are the output points, c's are the M coefficients of one row of the coefficient matrix and x's are the M elements of the input data vector. Since the coefficient structure is two-dimensional two types of scheduling functions are required to describe the basic computation in this case. The computation is performed according to the scheduling functions f(i), g's(i,j), i=0,1,2,...N-1, j=0,1,2,...M-1. The scheduling functions f(i), g's(i,j) are one-by-one and onto and in the original case f(i)=i and g's(i,j)=j. The f function determines the order in which the inner products are computed (order in which output terms are evaluated). The g's functions (one per row of the coefficient matrix) determine the order in which the partial products that form the inner products are computed and are different for each inner product. The total power cost (PC) of the sequence of evaluation of the partial products that constitute the basic computation as determined by the scheduling functions f, g's is given by the following equation:

$$PC(f,g) = \sum_{f(i)=0}^{N-1} \Big(\sum_{g(i,j)=0}^{M-1} PC(p_{f(i),g(i,j)}, p_{f(i),g(i,j)+1}) + PC(p_{f(i),N-1}, p_{f(i)+1,0}) \Big) \tag{7.8}$$

where $p_{f(i),g(i,j)}$ is the partial product $c_{f(i)g(i,j)}x_{g(i,j)}$. The power cost for a pair of partial products is given by the power related cost function selected to drive the synthesis procedure. The term $PC(p_{f(i),N-1}, p_{f(i)+1,0})$ denotes the power cost between the last partial product of the inner product f(i) and the first partial product of the inner product f(i)+1. This term is included because after the computation of the last partial product $p_{f(i),N-1}$ for the output term $y_{f(i)}$ the first partial product $p_{f(i)+1,0}$ for the term $y_{f(i)+1}$ must be computed. The proposed techniques aim at deriving the scheduling functions f, g's, that minimize the total power cost of the basic computation.

The formulation of the instruction-level scheduling problem is more complex for the transformational algorithms. In a first step the minimum cost scheduling of the inner products IP_i (i=0,1,...N-1) of the basic computation must be determined. In the second step the minimum cost scheduling of the partial products p_{ij} (j=0,1,...M-1) that constitute each inner product, must be found. The problem is tackled in the following way: For all possible pairs of inner products IP_i, IP_j, (i=0,1,...N-1, j=0,1,...N-1) the minimum power cost connection is determined. The minimum power cost connection is defined as the pair of partial products p_{ik} (belonging to the inner product IP_i, k=0,1,...N-1) and p_{jl} (belonging to the inner product IP_j, l=0,1,...N-1) that minimizes the power cost between all possible pairs of partial products as described by the following equation:

$$\begin{aligned} &minimum \quad cost \quad connection \quad for \quad IP_i, IP_j : (p_{i,k}, p_{j,l}) \\ &i,j = 0,1,\ldots N-1 \quad k,l = 0,1,\ldots M-1 \\ &PC(p_{ik}, p_{jl}) \leq PC(p_{im}, p_{jn}) \\ &\forall \quad m,n \quad m,n = 0,1,\ldots M-1 \end{aligned} \tag{7.9}$$

The complexity of the minimum cost determination procedure is $M^2N(N-1)$. Since the minimum power cost connection has been determined for all possible pairs of inner products the graph G (V, E) of the problem can be constructed where V is the set of N vertices corresponding to the N inner products while E is the set of edges that model the unconstrained transitions from one inner product to another. The graph is complete. To each edge the power cost of the minimum cost connection corresponding to the inner products connected by the edge is assigned as a cost. The determination of a minimum cost schedule of the inner products is formulated as a Traveling Salesman's Problem (TSP) on this graph. A closed path over all vertices (inner products) must be found, without passing from a vertex more than once, resulting to a minimum cost.

As soon as the minimum power cost scheduling of the inner products is determined, the partial products that constitute each inner product must be scheduled. Each vertex (inner product) of the initial problem's graph is considered as a graph with vertices corresponding to the partial products that constitute the inner product represented by the initial vertex. The power costs between the partial products are assigned as costs to the edges of the inner product graphs. However the scheduling of the inner products performed in the first step results in the determination of the partial products that will be evaluated first and last for each inner product. Thus the derivation of the minimum cost schedule of the partial products of an inner product can be formulated as a restricted minimum cost open path problem (over the vertices of the inner product graph). The problem is restricted since the starting and ending vertices of the path are determined by the inner products scheduling. This problem can be solved by first solving the corresponding Traveling Salesman's Problem on the graph consisting from the partial products of a given inner product and then removing the edge connecting the vertices corresponding to the partial products scheduled first and the last (determined by the inner product scheduling).

Transformational algorithms - Parallel output computation

The parallel computation of the output terms of a transformation of length $N \times M$ is described by the following equation, for j=0,1,...,M-1 and for all output points $y_{f(i,j)}$,

$$y_{f(i,j)} = c_{f(i,j),g(j)} \cdot x_{g(j)} \tag{7.10}$$

where y's are the output points, c's are the coefficients and x's are the elements of the input data vector. When the output terms of the basic computation of the transformational algorithms are computed in parallel each input data term is accessed only once from the memory element where it is stored and its contribution to all the output terms is computed. The computation is performed according to the ordering functions f's, g. The ordering functions f's, g are one-by-one and onto and in the original case f's(i,j)=i, g(j)=j. The f's functions determine the schedule in which the partial products (using a specific input data term) of the output terms are computed. The g function determines the schedule in which the input data are read from the memory (and the sets of partial products using the same data are computed). The total power cost (PC) of the sequence of evaluation of the $N \times M$ partial products required for the N output terms as determined by the ordering functions f's, g, is given by the following equation:

$$PC(f,g) = \sum_{g(j)=0}^{M-1} \Big(\sum_{f(i,j)=0}^{N-1} PC(p_{f(i,j),g(j)}, p_{f(i,j)+1,g(j)}) + PC(p_{N-1,g(j)}, p_{0,g(j)+1}) \Big) \tag{7.11}$$

where $p_{f(i,j),g(j)}$ is the partial product $c_{f(i,j)g(j)}x_{g(j)}$. The term $PC(p_{N-1,g(j)}, p_{0,g(j)+1})$ denotes the power cost between the last partial product of the column g(j) and the first partial product of the column g(j)+1. This term is included because after the computation of the last partial product $p_{N-1,g(j)}$ for the output term y_{N-1} the first partial product $p_{0,g(j)+1}$ for the term y_0 must be computed. The aim of the proposed techniques is to derive the ordering functions f's, g that minimize the total power cost (PC). The formulation of the instruction-level scheduling problem in this case is similar to the formulation of the sequential output computation case. The inner products of the sequential output computation case are replaced by the sets of partial products using the same input data term.

7.3.2.2 Multiple functional units case. The synthesis techniques described in the previous subsection assume that each basic computation is realized on a single functional unit (multiply/accumulator). The synthesis techniques that will be presented in this sub-section handle the cases, where basic computations are realized on more than one functional units.

Convolutional algorithms

In the general case, a basic convolution computation of this category can be easily implemented on K functional units by randomly (or sequentially with respect to the initial algorithm's definition) assigning the partial products to the functional units. In most cases this is not a good solution in terms of power consumption. The proposed techniques consist of the following two steps: a) Instruction-level assignment: Each of the N partial products is assigned to one of the K functional units available. The only restriction is the load balancing among the functional units. The N partial products must be clustered to K groups of highly correlated partial products leading to power cost reduction. For this reason the basic computation is scheduled first assuming realization on a single functional unit and then a clustering of the partial products is performed sequentially, in the same way as shown in section 7.3.1. b)Instruction-level scheduling: In this step, a schedule is derived for the partial products assigned to each functional unit. The single functional unit scheduling technique described in the previous subsection is used for this reason.

Transformational algorithms

It is assumed that no splitting of the basic computation's inner products to different functional units (MACs) is performed and that load balancing is preserved. Implementation of a basic computation in K functional units creates K sub-basic computations. Each of these sub-basic computations can be assigned to one functional unit. Two main cases exist:

a) K=N. The number of the available functional units is equal to the number of inner products. The instruction-level assignment is straightforward and one inner product is assigned to each functional unit. The scheduling of the partial products in each functional unit is performed in the same way as for the convolutional algorithms in the single functional unit case.

b) K<N. More than one inner products must be assigned to each functional unit. The proposed techniques consist of the following two steps: a) Instruction-level assignment: In this step groups of correlated inner products are created to reduce the power

cost, when passing from one inner product computation to another. The heuristic technique presented in section 7.3.1 is used for this reason. The inner products are first scheduled assuming single functional unit implementation. The resulting sequence of inner products is then divided sequentially to K groups of N/K inner products. Each of the groups is assigned to each functional unit. b) Instruction-level scheduling: It is assumed that the inner products assigned to each functional unit are computed sequentially. Thus the instruction-level assignment substep produces a schedule for the inner products assigned to the same functional unit as well. The task of the second step is to schedule the partial products within each inner product. This is achieved by using the same approach as for the single functional unit implementation of transformational algorithms.

7.4 COST FUNCTIONS

The synthesis techniques presented earlier require a power related cost function (PC) to drive the optimization procedure. This cost function may attempt to capture different parts of or the total power consumption in a realization of a given algorithmic kernel. Two main power-related cost functions are proposed:

Interconnect power cost function

This cost function captures the power consumed in the buses connecting the storage elements where the data and coefficients are stored to the functional units. Assuming two partial products $c_i d_i$ and $c_j d_j$ the cost function is given by the following equation:

$$PC_B(c_i d_i, c_j d_j) = CW_c \times HD(c_i, c_j) + CW_d \times AHD(d_i, d_j) \tag{7.12}$$

where CW_c , CW_d are weights related to the capacitances of the coefficient and data buses, $HD(c_i, c_j)$ represents the Hamming distance between the coefficients of the partial products that can be directly evaluated since coefficients are known before realization and $AHD(d_i, d_j)$ is the Average Hamming distance between the data terms of the partial products. The evaluation of the exact Hamming distance between two data samples is not possible since data are not known before run-time. An approximation (AHD) can be obtained by simulation of typical data of a specific application (e.g. test images for image processing applications or test sequences for video processing applications). In the case where the capacitance weights of the coefficient and data buses are equal the cost function can be simplified to the one given by equation 7.13:

$$PC_B(c_i d_i, c_j d_j) = HD(c_i, c_j) + AHD(d_i, d_j) \tag{7.13}$$

The above mentioned interconnect oriented power cost function includes both static and dynamic components. The interconnect oriented power cost function may become fully static by taking into consideration only the coefficients. In such a case the cost function is given by equation 7.14:

$$PC_B(c_i d_i, c_j d_j) = HD(c_i, c_j) \tag{7.14}$$

It must be noted that in deep submicron technologies the power consumed in the interconnect buses becomes more significant than the power consumed in the functional units.

Functional unit power cost function

This cost function approximates the power consumed in the functional units. It is more useful in cases where the interconnect length is small thus leading to small interconnect power. Assuming two partial products $c_i d_i$ and $c_j d_j$ the cost function is given by equation 7.15:

$$PC_I(c_i d_i, c_j d_j) = \sum_{\forall k} a_k \times c_k \tag{7.15}$$

where a_k is the switching activity in the k_{th} node of the functional unit and c_k is the node capacitance. In order to calculate the functional units related power weights, the method presented in [25] is used. This method is an incremental static (or probabilistic) power estimation approach, at the gate level of abstraction, that estimates the switching activity of a logic circuit under a real delay gate model. Capacitance weights are estimated based on the fan-out of the nodes. The chosen gate model allows accurate estimation of the functional and spurious (glitches) transitions, leading to accurate power estimation. The proposed approach manipulates efficiently large circuits employing a new partitioning heuristic, which estimates the switching activity with reduced computational complexity. The proposed method takes into consideration: a) first order temporal correlations, b) spatial correlations due to internal dependencies (Reconvergent Fanout Regions), c) simultaneous input switching and d) real delay gate model for the standard cells of a circuit. For each weight the input statistics are extracted from the corresponding vector set of the input samples and the algorithm's coefficients. These statistics are propagated through the circuit nodes, using the method [25] and hence the corresponding power weight is estimated. Each weight exhibits an acceptable accuracy in small computational time. The mean error in the power consumption that estimated is in the range of 5-10 compared with the power consumption that is calculated using QuickSim II gate level simulator of Mentor Graphics. The static nature of the coefficients of the algorithms under consideration leads even to increased accuracy of the proposed estimation technique.

The two power costs described above can be also combined to form a single cost function. The power related to the address busses of the data and coefficient storage elements can be easily incorporated in the cost function.

7.5 EXPERIMENTAL RESULTS

The proposed synthesis techniques have been implemented in ANSI C code. The prototype software may run either on UNIX or on Windows platforms. The prototype software has been used for the synthesis of sum-of-products based DSP kernels. In this section the effect of the proposed techniques on the power consumption of several typical DSP kernels is presented.

First the effect of the proposed synthesis techniques on the interconnect power is explored. The cost function given by equation 7.13 is used for this exploration. In this way the interconnect power is approximated by the switching activity at the data and coefficient buses assuming that the data and coefficient storage elements are connected to the functional units through bus lines of the same capacitance. Bit true simulations of ANSI C codes of the DSP kernels (using special instrumentation code as well) have been used to monitor the switching activities at the buses of realizations of the DSP kernels. The input buses switching activities obtained when the proposed synthesis techniques are applied, are compared to those obtained when the algorithms' computations are executed according to the algorithms' definitions and the related savings are computed.

The input buses switching activity savings achieved for different coefficient widths for two's complement representation are presented in figure 7.4. These results assume single functional unit realizations. The data that have been used for the simulation of the different algorithms are given in table 7.1. The fast DCT algorithm is described in [1]. The Fourier transforms are described in [24]. The 15, 11 and 9 tap filters are described in [23] while the 8 and 6 tap filters are described in [1]. In many cases savings of more than 25% are achieved. The larger savings are achieved for the DFT and 15-tap FIR filter while the smaller savings are achieved for the 6-tap and 8-tap FIR filters. An important observation is that the savings achieved do not increase proportionally to the coefficient bitwidth.

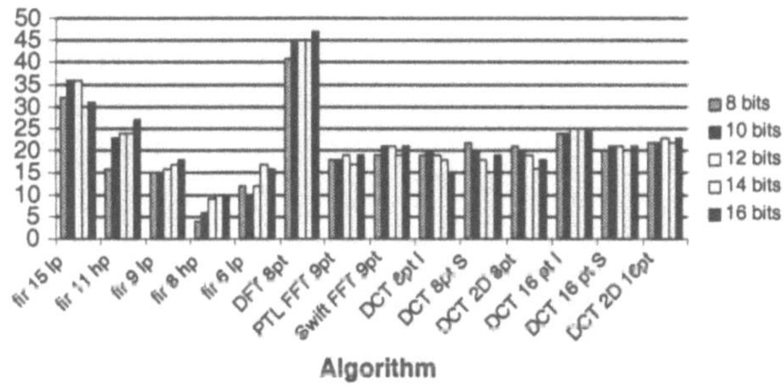

Figure 7.4 Interconnect power savings (in %) for different coefficient bitwidths and single functional unit realizations for two's complement representation

The results presented for the transformational algorithms correspond to the sequential output computation case. Results for the parallel output computation case are presented in figure 7.5. The savings achieved for the parallel output computation case are larger in comparison to those achieved for the sequential output computation case. This is because in the parallel output computation case no penalties in the data switching activity are introduced by the proposed techniques in comparison to the reference case.

The effect of the presence of multiple functional units on the results of the proposed synthesis techniques is illustrated in figure 7.6. Two's complement representation is assumed in this case as well. The data that have been used for simulation are described in table 7.2. The power savings achieved by the application of the proposed synthesis techniques do not decrease with the increase of the number of functional units.

Table 7.1 Simulation data for the single functional unit case

Algorithm	*Data*
FIR low pass 15 tap	Random 16 bits
FIR high pass 11 tap	Image 9 bits
FIR low pass 9 tap	Wavelet transform output (third level) 16 bits
FIR high pass 8 tap	Wavelet transform output (third level) 16 bits
FIR high pass 6 tap	Image 9 bits
DFT 1D 8pt	Image 9 bits
PTL FFT 1D 9pt	Image (preprocessed) 10 bits
Swift FFT 1D 9pt	Image (preprocessed) 10 bits
(Fast) 1D DCT 8pt I	Image (preprocessed) 10 bits
(Fast) 1D DCT 8pt S	Speech 16 bits
(Fast) 2D DCT 8pt	Image (preprocessed) 10 bits
(Fast) 1D DCT 16pt I	Image (preprocessed) 10 bits
(Fast) 1D DCT 16pt S	Speech 16 bits
(Fast) 2D DCT 16pt	Image (preprocessed) 10 bits

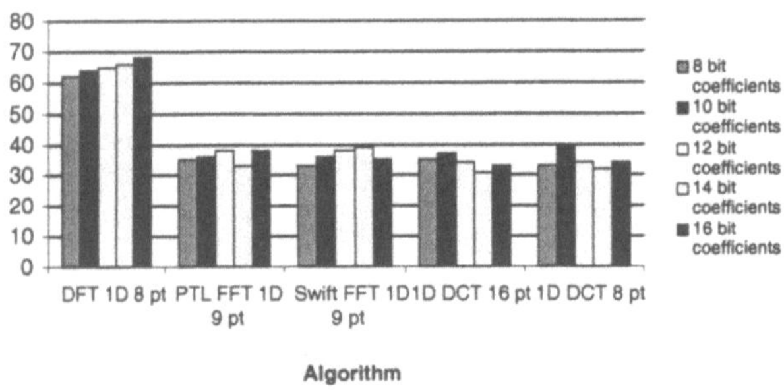

Figure 7.5 Interconnect power savings (in %) for transformational algorithms and the parallel output computation case

Two cost functions can be used to drive the proposed synthesis techniques as mentioned in section 7.4 A dynamic one (requiring simulation for its evaluation) given by equation 7.13 and a fully static one given by equation 7.14. Comparisons of the results achieved by the two different cost functions are presented in figure 7.7. The data used for the simulations are described in table 7.3. In all cases the differences in the savings achieved by the two different cost functions are small. This is important since no simulation is required for the evaluation of the fully static cost function speeding up significantly the synthesis process.

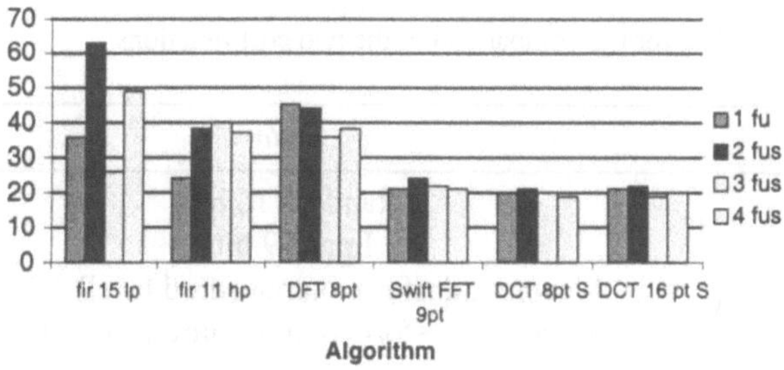

Figure 7.6 Interconnect power savings (in %) for multiple functional unit realizations and two's complement representation

Table 7.2 Simulation data for the multiple functional units case

Algorithm	*Data*	*Coefficient bitwidth*
FIR low pass 15 tap	Random 16 bits	12
FIR high pass 11 tap	Image 9 bits	14
DFT 1D 8pt	Image 9 bits	12
Swift FFT 1D 9pt	Image (preprocessed) 10 bits	12
Fast 1D DCT 8pt	Speech 16 bits	10
Fast 1D DCT 16pt	Speech 16 bits	10

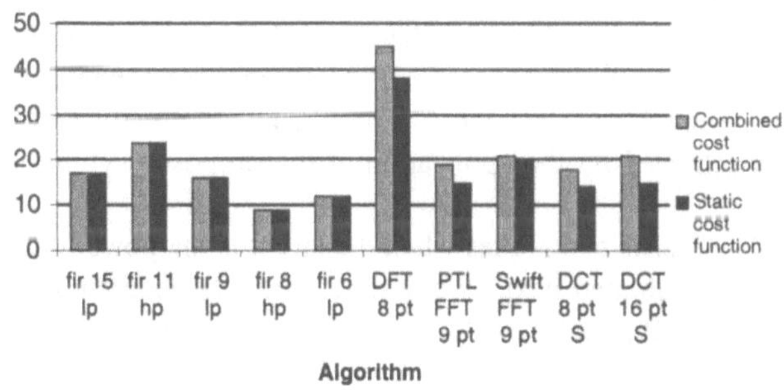

Figure 7.7 Comparison of the two cost functions (in absolute numbers) for two's complement representation

Experimental results from the application of the proposed synthesis techniques to a complex algorithm namely the two-dimensional three level discrete wavelet transform

Table 7.3 Simulation data for the comparison of the two cost functions

Algorithm	*Data*
FIR low pass 15 tap	Random 16 bits
FIR high pass 11 tap	Image 9 bits
FIR low pass 9 tap	Wavelet transform output (third level) 16 bits
FIR high pass 8 tap	Wavelet transform output (third level) 16 bits
FIR high pass 6 tap	Image 9 bits
DFT 1D 8pt	Image 9 bits
PTL FFT 1D 9pt	Image (preprocessed) 10 bits
Swift FFT 1D 9pt	Image (preprocessed) 10 bits
(Fast) 1D DCT 8pt S	Speech 16 bits
(Fast) 1D DCT 16pt	Speech 16 bits

that requires evaluation of several basic computations are presented in table 7.4. It is assumed that each basic computation is evaluated on a single multiply-accumulator. For the coefficients, a 16-bit representation is assumed. The input data are represented by 8 bits while for the representation of the wavelet coefficients 16 bits are used. Six different filter sets have been used and simulated. The EPIC filters are described in [23], the Daubechies and Johnston's filters are described in [1]. The switching activity in the coefficient bus is significantly reduced. The proposed techniques introduce penalties (worst case 12%) in the switching activity in the data bus because data correlation is destroyed. Specifically the reordering of computation introduces switching activity penalties in the inputs of the three levels of the wavelet decomposition because at these points the correlation of data is high. In the intermediate and output stages of the decomposition levels data are less correlated and thus the result of the reordering is more random (either small savings or small switching activity penalties are introduced). However, the penalties in data activity introduced are much smaller in comparison to the savings achieved by the application of the proposed synthesis techniques.

The effect of the proposed synthesis techniques when the target is the power consumed in the functional units is evaluated next. Word serial multiply-accumulators based on two different types of multipliers namely a Carry Save (CSA) array multiplier and a Booth encoded-Wallace multiplier have been developed. The VHDL descriptions for the multipliers have been obtained using the DesignWare software of Synopsys running on a HP-10 workstation. The functional units power has been estimated through switch level simulations and by combining the activity results produced with the capacitance files corresponding to the target technology ($0.7\mu m$). The estimation procedure has been carried out using the Mentor Graphics framework running on a HP-10 workstation. The proposed techniques have been applied to a number of DSP kernels. The data used for the simulations, the data and coefficient bitwidths and the number of functional units used for the realizations of the DSP kernels are presented in table 7.5. The results of the application of the proposed synthesis techniques

Table 7.4 Simulation results for the 2-D wavelet transforms

Filter set	*Change in coefficient activity #transitions(%)*	*Change in data activity #transitions(%)*	*% Change in total input activity*
EPIC-15 tap	-20300019(-51.8%)	+2698508(+12%)	-29
EPIC-11 tap	-13418613(-48.1%)	+954075(+5.4%)	-28
EPIC-9 tap	-8945772(-43.3%)	+853493(+6.5%)	-24
Daubechies-6 tap	-2408421(-14%)	+281648(+3.1%)	-9
Daubechies-8 tap	-2752521(-13.3%)	+646421(+5.4%)	-9.5
Johnston's-8 tap	-5849097(-34.3%)	+599234(+5.1%)	-17

on the functional units power consumption are presented in table 7.6. Power figures are in μW and have been obtained assuming a frequency of operation of 20 MHz, and a supply voltage of 5 volts. Two's complement representation is assumed as well. Power savings of 15% on average are achieved by the application of the proposed techniques. As far as the multiplier architecture is concerned the proposed synthesis techniques lead to larger savings in the case of the Carry Save array multiplier.

Table 7.5 Demonstration algorithms and simulation data

Algorithm	*Data*	*Coefficient bitwidth*	*Number of total functional units*
FIR 11 hp	image 9 bits	10	1
FIR 9 lp	image 9 bits	10	1
FIR 8 lp	image 9 bits	10	1
PTL FFT 9	image preprocessed 10 bits	12	1
Swift FFT 9	image preprocessed 10 bits	12	1
DCT 8 (1)	image preprocessed 10 bits	12	1
DCT 8 (2)	image preprocessed 10 bits	10	2
DCT 8 (3)	image preprocessed 10 bits	10	4
DCT 8 (4)	1D DCT output (on image data) 16 bits	16	1
DCT 8 (5)	1D DCT output (on image data) 16 bits	16	2
DCT 8 (6)	1D DCT output (on image data) 16 bits	16	4

Table 7.6 Functional unit power reduction for two's complement representation

Algorithm	*Initial power Array-CSA*	*% reduction Array-CSA*	*Initial power Wallace-Booth*	*% reduction Wallace-Booth*
FIR 11 hp	10505	22,9	12604	27,9
FIR 9 lp	9225	19,2	12095	22,5
FIR 8 lp	10897	2,0	10065	20,9
PTL FFT 9	17312	12,8	14267	6,3
Swift FFT 9	19300	33,7	14982	7,3
DCT 8 (1)	12517	11,1	14114	6,6
DCT 8 (2)	4620	15,4	6026	7,7
DCT 8 (3)	9129	12,7	11734	7,9
DCT 8 (4)	45481	5,8	33762	5,8
DCT 8 (5)	17474	16	16064	6,9
DCT 8 (6)	34953	32,8	31886	43,6
Average	22428	16,8	17493	14,9

7.6 COMPARISON WITH EXISTING APPROACHES

Most of the existing power conscious instruction-level scheduling and assignment techniques [3], [7], [14], [19], [20], [22] mainly aim at minimizing the power consumption of the functional units. However the power consumption of the interconnect buses, which is targeted by the proposed techniques, becomes more important especially in the deep submicron technologies. All the techniques described in [7], [14], [19], [20], [22] use time consuming simulations to evaluate their cost functions. No fully or partly static cost functions are proposed by the techniques [7], [14], [19], [20], [22], which is the case in the proposed synthesis techniques. The idea of the exploitation of the presence of operations with common input operands (sibling operations) proposed in [7] is implicitly included in the proposed synthesis techniques. Operations with input operands closely matched at the bit-level (including the case of identical input operands) will be scheduled at successive control steps by the proposed techniques. The techniques described in [3], [7], [19], [20], [22] use complex techniques to formulate the power efficient synthesis procedure. The techniques proposed in this paper taking advantage of the independent nature of the sum-of-products computations use a very simple and easy to encode formulation for the synthesis tasks using the Traveling Salesman's Problem. This problem can be solved very efficiently. No systematic formulation is proposed for the application of the techniques described in [14]. The technique for simultaneous scheduling and binding proposed in [5] aims at reducing the power consumed in the interconnect buses. This technique is based solely on time consuming simulations to evaluate the cost function while the proposed techniques use either partly or fully static cost functions to reduce the time of the synthesis procedure. Furthermore the synthesis technique described in [5] uses an iterative improvement approach starting from a random solution. Such an approach requires longer run times than the proposed heuristics to produce a good quality solution. The techniques described in [9] and [13] can be used as a pre-processing step for the proposed synthesis techniques.

The only existing synthesis technique aiming explicitly at reducing the power consumption in realizations of sum-of-product computations is proposed in [2]. This synthesis strategy proposes schedules for the evaluation of the partial products of convolutional type of computations based on the values of the coefficients. The derived schedules lead to reduction of the switching activity at the nodes of customized realizations of the inner product computations. Techniques targeting explicitly transformational type algorithms with two dimensional coefficient structures are not proposed in [2]. The proposed techniques are compared to those presented in [2] in terms of the switching activity in the buses connecting the functional units to the storage elements of the coefficients and the data. For the techniques presented in [2] it is assumed that the coefficients (and the partial products) of the inner products are ordered to satisfy the relation $c_0 > c_1 > .... > c_{N-1}$, where c_i is the coefficient of the partial product scheduled i_{th}. Ordering satisfying the relation $c_0 < c_1 < .... < c_{N-1}$, is expected to lead to similar results as stated in [2]. For the transformational algorithms the inner products are assumed to be executed in the order determined by the original algorithm definition. The algorithms and the parameters used for the simulation are presented in table 7.7. Single functional unit realizations are assumed in all cases. The results of the comparison are presented in table 7.8. The results prove that the proposed techniques outperform the techniques presented in [2] in all cases. In average the proposed techniques lead to 5% improvement in terms of input buses switching activity over the techniques proposed in [2].

Table 7.7 Algorithms and simulation data for the comparison in terms of bus switching activity of the proposed techniques and the techniques described in [2]

Algorithm	*Data*	*Coefficient bitwidth*
FIR 11 hp	image 9 bits	10
Swift FFT 9	image 10 bits	12
1D Fast DCT 16 (1)	image 10 bits	12
1D Fast DCT 16 (2)	Output of 1D DCT 16 bits	16
1D Fast DCT 8 (1)	image 10 bits	12
1D Fast DCT 8 (2)	Output of 1D DCT 16 bits	16

A large number of techniques have been proposed for the power optimization of sum-of-products computation based digital signal processing algorithms. Many of these techniques are based on the modification of the coefficients to reduce power consumption. The techniques described in [15], [6], [8], [17], [21], transform the values of the algorithms' coefficients to reduce the complexity of the required multiplications. The proposed synthesis techniques can be applied after the optimization of the algorithms' coefficient structures. As far as the reordering of computation is concerned the technique described in [12] perturbs the filter coefficients to decrease

Table 7.8 Comparison in terms of bus switching activity of the proposed techniques and the techniques described in [2]

Algorithm	*% reduction two's complement proposed*	*% reduction two's complement [2]*	*% reduction sign magnitude proposed*	*% reduction sign magnitude [2]*
FIR 11 hp	25	21,9	25	21,9
Swift FFT 9	21,3	16,4	20,8	16,7
1D Fast DCT 16 (1)	25,4	17,9	25,4	17,9
1D Fast DCT 16 (2)	21,5	17,8	23	18,7
1D Fast DCT 16 (3)	19,5	10,9	19,5	10,9
1D Fast DCT 16 (4)	19,4	15,8	17,1	15
Average	22	16,8	21,8	16,9

the number of operations required to obtain the filter output (and not the switching activity which is the case for the proposed synthesis techniques) while preserving the desired filter characteristics. In [10] the parallel output computation case is proposed for transformational type algorithms in order to keep the data input operand stable for successive partial products and thus to reduce power consumption. The proposed techniques enhance further the parallel output computation technique by reordering the partial products that use the same data input in order to minimize the switching activity in the coefficient input buses as well.

7.7 CONCLUSIONS

Novel data path synthesis techniques for sum-of-products computations have been presented. The proposed techniques take advantage of the special characteristics of sum-of-products computations. Simple and efficient instruction-level scheduling and assignment heuristics based on the Traveling Salesman's Problem have been described. Cost functions targeting different components of the total power dissipation that can be partly statically evaluated have been proposed. Experimental results have proven that the proposed techniques lead to 25% savings in terms of the interconnect power and to 15% savings in terms of functional units power on average in comparison to reference implementations.

References

[1] V. Bhaskaran, K. Konstantinides, Image and Video Compression Standards, Kluwer Academic Publishers, 1994.

[2] A. Chatterjee, R. Roy, "Synthesis of low power linear DSP circuits using activity metrics", International Conference VLSI Design, January 1994, pp. 265-270.

[3] J-M. Chang, M. Pedram, "Module assignment for low power", ACM/IEEE European Design Automation Conference, September 1996, pp. 376-381.

[4] N. Christofides, "Worst-case analysis of a new heuristic for the travelling salesman problem", Report 388, Graduate School of Industrial Administration, Carnegie Mellon University, Pittsburgh, PA.

[5] A. Dasgupta, R. Karri, "Simultaneous scheduling and binding for power minimization during microarchitecture synthesis", ACM/IEEE International Symposium Low Power Design, April 1995, pp. 69-74.

[6] A. G. Dempster and M. D. Mcleod, "Use of minimum adder multiplier blocks in FIR digital filters", IEEE Transactions on Circuits and Systems II, vol. 42, pp. 569-577, September 1995.

[7] Y. Fang, A. Albicki, "Joint scheduling and allocation for low power", IEEE International Symposium on Circuits and Systems, May 1996, pp. 556-559.

[8] R. Hartley, "Optimization of CSD multipliers for filter design", IEEE International Symposium on Circuits and Systems, 1991, pp. IV.1992-IV.1995.

[9] M. Janssen, F. Catthoor, H. De Man, "A specification invariant technique for regularity improvement between flow graph clusters", European Design Automation Conference, February 1996, pp. 138-143.

[10] H. Kojima, A. Shridhar, "Interlaced accumulation programming for low power DSP", ACM/IEEE International Symposium Low Power Electronics and Design, August 1996, pp. 213-216.

[11] M. T-C. Lee, V. Tiwari, S. Malik, M. Fujita, "Power analysis and minimization techniques for embedded DSP software", IEEE Transactions Very Large Scale Integration Systems, pp. 123-135, June 1997.

[12] M. Mehendale, S. D. Sherlekar, and G. Venkatesh, "Coefficient optimization for low power realization of FIR filters", IEEE Workshop VLSI Signal Processing, September 1995, pp. 352-361.

[13] M. Miranda, F. Catthoor, M. Janssen, H. De Man, "High-level address optimisation and synthesis techniques for data transfer intensive applications", IEEE Transactions Very Large Scale Integration Systems, vol.6, pp. 677- 686, December 1998.

[14] E. Musoll, J. Cortadella, "High-level synthesis techniques for reducing the activity of functional units", ACM/IEEE International Symposium Low Power Design, April 1995, pp. 99-104.

[15] H. Nguyen, A. Chatterjee, R. Roy, "Activity measures for fast relative power estimation in numerical transformation for low power DSP synthesis", Journal of

VLSI Signal Processing, Kluwer Academic Publishers, vol. 18, pp. 25-38, January 1998.

[16] C. H. Papadimitriou, K. Steiglitz, Combinatorial Optimisation: Algorithms and Complexity, Prentice-Hall, 1982.

[17] D. Pearson, K. Parhi, "Low power FIR digital filter architecture", IEEE International Symposium Circuits and Systems, April-May 1995, pp. 231-234.

[18] J. M. Rabaey, M. Pedram, Low Power Design Methodologies, Kluwer Academic Publishers, 1996.

[19] A. Raghunathan, N. Jha, "Behavioral synthesis for low power", International Conference Computer Design, October 1994, pp. 318-322.

[20] A. Raghunathan, N. Jha, "An ILP formulation for low power based on minimizing switched capacitance during data path allocation", IEEE International Symposium Circuits and Systems, April-May 1995, pp. 1069-1073.

[21] N. Sankarayya, K. Roy, D. Bhattacharya, "Algorithms for low power and high speed FIR filter realization using differential coefficients", IEEE Transactions Circuits and Systems II, vol. 44, pp. 488-497, June 1997.

[22] D. Shin, K. Choi, "Low power high level synthesis by increasing data correlation", ACM/IEEE International Symposium Low Power Electronics and Design, August 1997, pp. 62-67.

[23] E. P. Simoncelli, E. H. Adelson, "Subband transforms" Subband Image Coding, editor John Woods, Kluwer Academic Publishers, 1990.

[24] W. Smith, J. Smith, Handbook of Real-Time Fast Fourier Transforms, IEEE Press, 1995.

[25] S. Theoharis, G. Theodoridis, N. Zervas and C. Goutis, "Accurate and fast power estimation of large combinational circuits", IEEE Workshop Power and Timing Modeling, Optimization and Simulation, October 1999, pp 199-208.

Index

MIX
Papier aus verantwortungsvollen Quellen
Paper from responsible sources
FSC® C105338

If you have any concerns about our products,
you can contact us on
ProductSafety@springernature.com

In case Publisher is established outside the EU,
the EU authorized representative is:
Springer Nature Customer Service Center GmbH
Europaplatz 3, 69115 Heidelberg, Germany

Printed by Libri Plureos GmbH
in Hamburg, Germany